# David Levy's Guide to Eclipses, Transits, and Occultations

In this simple guide, David Levy inspires readers to experience the wonder of eclipses and other transient astronomical events for themselves. Covering both solar and lunar eclipses, he gives step-by-step instructions on how to observe and photograph eclipses. As well as explaining the science behind eclipses, the book also gives their historical background, discussing how they were observed in the past and what we have learned from them. This personal account contains examples from the 77 eclipses the author has witnessed himself.

The guide also includes chapters on occultations of stars and planets by the Moon and of asteroids by stars, and the transits of Mercury and Venus. Tables of future eclipses make this invaluable for anyone, from beginners to practiced observers, wanting to learn more about these fascinating events.

**David H. Levy** is President of the National Sharing the Sky Foundation, and is one of the most successful comet discoverers in history. He has discovered 23 comets (eight of them using his own backyard telescopes) and was co-discoverer of Shoemaker–Levy 9, the comet that collided with Jupiter in 1994 producing the most spectacular explosions ever witnessed in the Solar System. Asteroid 3673 (Levy) was named in his honor. He has written three dozen books, is a contributing editor and monthly columnist for *Astronomy*, and was the former Science Editor for *Parade* magazine. In 1998 he won an Emmy as part of the writing team for the Discovery Channel documentary, *Three Minutes to Impact*. He now holds a Ph.D. from the Hebrew University of Jerusalem in the field of literature and the night sky.

# David Levy's Guide to Eclipses, Transits, and Occultations

DAVID H. LEVY

# CAMBRIDGE
## UNIVERSITY PRESS

University Printing House, Cambridge CB2 8BS, United Kingdom

One Liberty Plaza, 20th Floor, New York, NY 10006, USA

477 Williamstown Road, Port Melbourne, VIC 3207, Australia

314-321, 3rd Floor, Plot 3, Splendor Forum, Jasola District Centre, New Delhi - 110025, India

103 Penang Road, #05-06/07, Visioncrest Commercial, Singapore 238467

Cambridge University Press is part of the University of Cambridge.

It furthers the University's mission by disseminating knowledge in the pursuit of education, learning and research at the highest international levels of excellence.

www.cambridge.org
Information on this title: www.cambridge.org/9780521165518

First published 2010

*A catalogue record for this publication is available from the British Library*

*Library of Congress Cataloging in Publication data*
Levy, David H., 1948–
    David Levy's Guide to Eclipses, Transits, and Occultations / David H. Levy.
        p.   cm.
    Includes bibliographical references.
    ISBN 978-0-521-16551-8 (pbk.)
    1. Eclipses.   2. Transits.   3. Occultations.   4. Astronomy–Technique.
I. Title.   II. Title: Guide to eclipses, transits, and occultations.
    QB175.L48 2010
    523.7′8–dc22

                                                                    2010016792

ISBN  978-0-521-16551-8  Paperback

# Contents

# *Introduction*

Only the Earth doth stand forever still:
Her rocks remove not, nor her mountains meet;
(Although some wits enrich'd with learning's skill
Say heaven stands firm and that the Earth doth fleet
And swiftly turneth underneath their feet)
Yet, though the Earth is ever steadfast seen,
On her broad breast hath dancing ever been.

(Sir John Davies, *Orchestra, 1596*)

Whether day ended 400 years ago in Sir John Davies's time, or as it does today, the night sky that has attracted people for thousands of years begins another nightly show. For committed astronomers, amateur or professional, that darkening sky is all that is needed to get our juices flowing. Others require a little more, not just a static display that changes subtly from hour to hour, but something startling, something that crashes upon the celestial stage. A bright meteor, or an eclipse, can spark a lifelong interest in the sky. Eclipses are predictable, and there is usually nothing subtle about them. We can take the experience of an eclipse and put it into a bottle of fond memories. Eclipses show that the sky does change, that the sky is the show that never ends. Eclipses can inspire, and that is why I wrote this guide to getting the most from them.

## An eclipse journey

Eclipses are so interesting that some people travel the world to catch them. My wife Wendee and I did this quite literally in late 2003, when we flew

1

across the Atlantic from our home near Tucson, Arizona. Our first stop was London, the city of Shakespeare. Shakespeare knew eclipses, mentioning them and their implications several times in his writing. But our homage to Shakespeare was brief. We boarded another plane and landed half a day later deep in the southern hemisphere in Cape Town, South Africa. Even that city, famous as the place where, in the early nineteenth century, John Herschel and the Cape Observatory opened our vision of the southern sky, was but a stop on the way as we continued southwards down the line that divides the Indian from the south Atlantic oceans. We crossed the Antarctic Circle and still sped southwards, until our plane touched down on thousands of feet of ice. Finally, we were there: the Russian science station in the land of the midnight Sun.

But if the weather stayed clear, this would be no ordinary midnight Sun. The sky was the steadiest I had ever seen – the spots I saw through my 3.5-inch Questar telescope were so steady they looked like drawings. As the Sun began to set, we headed off across the ice to our viewing location and arrived there about 45 minutes later. Still the Sun was setting. We set up our telescopes and watched as the Sun touched the horizon. The Sun continued to set. An hour later, it was still setting. Then the tiniest nick appeared in its side. It was the Moon's first bite that would rapidly get bigger and culminate in total eclipse. Twenty minutes later, the Sun reached its lowest point on the horizon and began to rise again. With seven minutes left before the onset of totality, I noticed faint but definite shadow bands crossing the sunlit area of snow in front of us. They appeared as quite regular waves of dark lines moving away from the Sun at the rate of a meter per second. As totality drew nearer, these bands darkened and grew more obvious. I saw the same effect after totality, with the bands moving in the same direction – so the total shadow band viewing time was about 12 minutes.

At two minutes before totality, the Moon's umbral shadow approached rapidly from a point just to the northeast of the Sun. At this point I made a big mistake. In order to access the telescope's filter and also take photographs, I removed my gloves. By the time totality ended less than 3 minutes later, both thumbs and several fingers were mildly frostbitten and severely painful. But the consequences of that cold had to wait. For as I looked again toward the disappearing Sun I saw the strangest natural scene I've ever encountered. Ahead of me was an expanse of snow and ice which met the horizon where a huge inverted cone of darkness soared into the sky. Due to the low Sun angle

The May 20th, 1947 Brazilian Eclipse
Discovered at the National Archives II
By Charles Wacker
July 24, 2003

Presented to David Levy
October, 2003

**Figure 1** A total eclipse of the Sun seen from Brazil by F. Barrows Colton, May 20, 1947. Image courtesy Charles Wacker, July 24, 2003.

that shadow was darker than I had ever seen it. In contrast to the rapid partial phase, I now thought that the eclipse was creeping towards totality. Already darker than a usual eclipse, there was still a thinning crescent. Finally the corona appeared faintly, and with it an "emerald cut" long diamond ring. Within a few seconds the corona blossomed to its full brilliance. At one o'clock (going clockwise around the Sun) was a huge coronal flare.

Visualize what was going on over the field of ice that minute: the Sun was grazing the horizon – in fact a slice of the lower edge, or limb was actually still below the horizon. During the next minute the shadow at the horizon moved from south to north like the slide of some vast heavenly ruler. The corona, thanks to the Moon illusion that affects our perception of objects near the horizon, looked huge against the sky. It was that intersection of Sun, Moon, horizon, and ice that was the highlight of this eclipse.

**Figure 2** All wearing special eclipse protection glasses, a large crowd waits anxiously for totality to begin on the morning of August 11, 1999 in the northwestern Atlantic Ocean, southeast of Halifax, Nova Scotia, on the *Regal Empress*. The lady second from left is Patsy Tombaugh; Len Wallach is wearing a white sweater and his daughter Joan-Ellen is standing to his right. Wendee and I are in the back row. Photograph by Roy Bishop.

A second emerald cut diamond ring ended totality, and I swung round to the east to watch the shadow speed off across an expanse of flat ice that stretched to the other horizon. The shadow quickly raced across the ice, but for 1¼ minutes on that cold day in 2003, the shadow of the Moon encountered us as we stood atop of miles of thickness of ice. As a result, we saw a darkened Sun at midnight.

### Experiencing an eclipse

Whether it involves a barely perceptible shading of one edge of the Moon or a complete blocking of the Sun, an eclipse is an experience not easily forgotten. More than any other factor, an eclipse is a cosmic display of how our world interacts with its Moon and the Sun. It is a dance worthy of the sixteenth-century poet John Davies, who in 1596 wrote about the cosmic dance in *Orchestra,* a stanza of which adorns the opening page of this introduction.

However Davies might have been aware of this beautiful interplay among worlds, the idea of a cosmic dance is expressed perfectly during an eclipse. When my parents and I, along with friend Paul Astrof, saw the beginning of the partial phase of the total eclipse of July 20, 1963, Dad looked at his watch amazed that the eclipse began exactly at the predicted time. I'll never forget his gazing through filtered glass at the tiny nick in the Sun, then at his watch. Conceivably, the 1963 eclipse could have been predicted 4.5 billion years ago, shortly after the Moon was formed out of a collision between the Earth and another world the size of Mars.

This is the magic of eclipses. In the pages that follow, I will offer inspiration, information, and indications to point your way towards a more enriching eclipse experience.

# Part I THE MAGIC AND HISTORY OF ECLIPSES

# 1

## Shakespeare, King Lear, and the Great Eclipse of 1605

> These late eclipses of the sun and moon portend no good to us.
>
> (Shakespeare, *King Lear*, 1.2.101–102)

The shadow of the Moon dropped swiftly out of the sky, charging along the sunrise terminator at 12,000 miles an hour. As the Sun edged higher, that velocity slowed to a more manageable 2000 mph as it tore southeast across the north Atlantic, crossed just west of England, and made landfall in southern France and northern Spain. The date: October 12, 1605. I like to think the event was observed by people all over Europe, including groups of people standing by the Thames in central London. Far out of town, King James I was out probably enjoying a day of hunting. As the day progressed, few let the almost imperceptible onset of a solar eclipse interrupt their business. But as more of the Moon moved across the Sun, by noon the sky was darkening noticeably and rapidly. Between 12:40 and 1:00 pm the sky was a twilight dark. Through breaks in the clouds, the Sun appeared as a thin curved line of light.

As the Moon continued moving eastward across the Sun, it abandoned its grip just after 2:00 pm. I like to imagine a group of Londoners peering at the sky as the eclipse ended, discussing its meaning. "I heard about these eclipses," said one, "at the theatre a few weeks ago."

"King Lear," another nodded. "But weren't there two eclipses?" Indeed there were. Two weeks earlier, during the predawn hours of September 27, a deep partial eclipse of the Moon darkened the predawn sky over London. Had any of those Londoners seen the lunar eclipse also, they would be part of a rare group of people who have seen eclipses of the Moon, and of the Sun,

**Figure 1.1** This is the home in which William Shakespeare was born and spent his early years. I like to imagine James Joyce's scenario in *Ulysses* that as a young boy of 9:

*A star, a daystar, a firedrake rose at his birth. It shone by day in the heavens alone, brighter than Venus in the night, and by night it shone over delta in Cassiopeia, the recumbrant constellation which is the signature of [Shakespeare's] initial among the stars. His eyes watched it, lowlying on the horizon, eastward of the bear, [but westward from Polaris] as he walked by the slumberous summer fields at midnight* – writing about the supernova in 1572.

(James Joyce, *Ulysses*, 1922)

— Although this story is apocryphal, it is difficult to fathom that the young Shakespeare, with his eclectic interests and zest, could have possibly missed seeing the great star of 1572.[1] Photo by David H. Levy.

separated by only two weeks. I often ask that question when I give lectures about eclipses, and even among experienced watchers of the sky, few have seen such a pair of eclipses from the same location.

Shakespeare might be in that group; at the least he knew that eclipses could happen in pairs. In *King Lear*, he describes a conversation about eclipses. The eclipse passages are a seminal discussion about the role that these transient events play, or do not play, in our lives. It begins with an argument by the

**Figure 1.2** This is the building likely used as Shakespeare's school. In the sixteenth century, towns in England tended to place a substantial emphasis on the education of their children, and Stratford had one of the best schools in that part of England, complete with teachers trained at Oxford and Cambridge.

Earl of Gloucester: "These late eclipses of the sun and moon portend no good to us. Though the wisdom of nature can reason it thus and thus, yet nature finds itself scourged by the sequent effects. Love cools, friendship falls off, brothers divide."[2]

Not so, counters his son Edmund, who fires back: "This is the excellent foppery of the world, that when we are sick in fortune, often the surfeits of our own behavior, we make guilty of our disasters the sun, the moon, and stars; as if we were villains on necessity . . ."[3]

Besides the two eclipses in the autumn of 1605, a total lunar eclipse darkened the full Moon that April. But it was the eclipse of the Sun that attracted the most attention. King James I himself wrote of it in a letter to Robert Cecil, his friend and aide. In it he pokes fun at those who would attach astrological implications to eclipses: "But now I will go to higher matters and tell you what I have observed . . . the effects of this late eclipse for as the troglodytes of the Nile that dwelt in caverns, the shepherds of Arcadia dwelling in little cabins, the Tartars harboring in their tents like the old patriarchs, so I, having now remained awhile in this hunting cottage, am abler to judge

**Figure 1.3** Shakespeare's Holy Trinity Church is on the west shore of the River Avon. The supernova of 1572 could have been seen toward the north, off to the right of the photograph. Photograph by David H. Levy.

of astronomical motions than ye who lives in the delicious courts of princes. The effects then of this eclipse for this year are very many and wondrous . . ."[4]

### The magic of eclipses

Shakespeare was taking advantage of the eclipses to engage his audience's interest in the narrative of *King Lear*. He uses the effects of eclipses often in his writing, but in this particular play eclipses are part of the great cosmic picture of darkness and storm that Shakespeare is using to discover a relation between humanity and the cosmos. Just as Shakespeare used the eclipses of 1605 to arouse the curiosity of his audience, this book aims to use eclipses as tools to inspire a deeper interest in the sky. In October of 2005, exactly 400 years after the great eclipse, I stood in a square in the city of Barcelona. Under a mostly cloudy sky, the Sun did peek through at 2 pm, the time that it would have been in the total phase of the eclipse. My mind harked back to that distant hour as the Moon's shadow crossed over Europe. I imagined people looking up at the Sun, admiring, wondering, and inquiring.

# Total Solar Eclipse of 1605 Oct 12

Geocentric Conjunction = 12:32:26.3 UT     J.D. = 2307559.022527
Greatest Eclipse = 12:57:42.8 UT     J.D. = 2307559.040078

Eclipse Magnitude = 1.03436     Gamma = 0.80195

Saros Series = 137     Member = 13 of 70

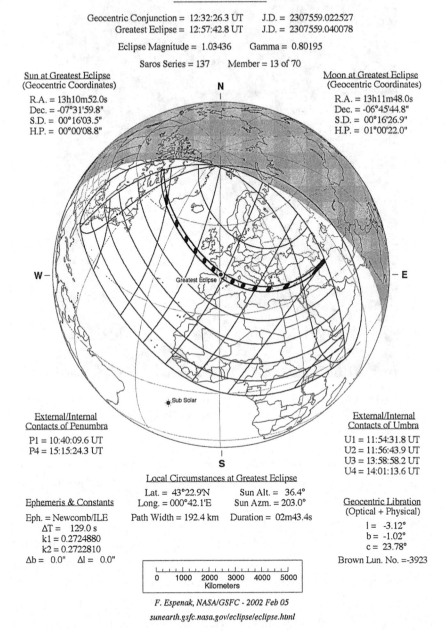

Sun at Greatest Eclipse
(Geocentric Coordinates)

R.A. = 13h10m52.0s
Dec. = -07°31'59.8"
S.D. = 00°16'03.5"
H.P. = 00°00'08.8"

Moon at Greatest Eclipse
(Geocentric Coordinates)

R.A. = 13h11m48.0s
Dec. = -06°45'44.8"
S.D. = 00°16'26.9"
H.P. = 01°00'22.0"

N

W —                                                — E

Greatest Eclipse

Sub Solar

S

External/Internal
Contacts of Penumbra

P1 = 10:40:09.6 UT
P4 = 15:15:24.3 UT

External/Internal
Contacts of Umbra

U1 = 11:54:31.8 UT
U2 = 11:56:43.9 UT
U3 = 13:58:58.2 UT
U4 = 14:01:13.6 UT

Local Circumstances at Greatest Eclipse

Lat. = 43°22.9'N     Sun Alt. = 36.4°
Long. = 000°42.1'E     Sun Azm. = 203.0°
Path Width = 192.4 km     Duration = 02m43.4s

Ephemeris & Constants

Eph. = Newcomb/ILE
ΔT = 129.0 s
k1 = 0.2724880
k2 = 0.2722810
Δb = 0.0"    Δl = 0.0"

Geocentric Libration
(Optical + Physical)

l = -3.12°
b = -1.02°
c = 23.78°

Brown Lun. No. =-3923

```
|_|_|_|_|_|_|_|_|_|_|_|_|
0   1000  2000  3000  4000  5000
          Kilometers
```

F. Espenak, NASA/GSFC - 2002 Feb 05
sunearth.gsfc.nasa.gov/eclipse/eclipse.html

**Figure 1.4** The path of Shakespeare's eclipse. Fred Espenak, from NASA's Goddard Space Flight Center, kindly provided this map of the path of the October 12 (N.S. = October 2 O.S.) 1605 total eclipse of the Sun, which tracked just to the south of England and whose track is produced with his kind permission. The map shows the path crossing the North Atlantic, and leaving all of England in a deep partial eclipse, then continuing through southern Europe and the northern part of the Middle East.

**Figure 1.5** The Galileo space probe snapped this image of the Earth and Moon family while passing by on its way to Jupiter. Would that Shakespeare could have seen this view!

As with any major eclipse, interest in astronomy would have increased as a result. But how was that interest maintained and built upon? In 1605, people watching *King Lear* might have remembered it. The 1605 eclipse also prompted Johannes Kepler to suggest that the corona of light around the Moon that day was actually the Moon's atmosphere. That idea might have caused people to think differently about the Moon. If the Moon has an atmosphere, could there be life there? We know differently now – the corona is the faint and hot atmosphere of the Sun, not the Moon – but one can imagine how the idea might have caused people to use the experience of an eclipse to conjure the possibility of life on other worlds.

The cosmos in 1605 seemed vastly different from the one we understand today. In earlier years John Dee, Queen Elizabeth's friend and informal science advisor, wrote that "The Sun, when he is farthest from the earth,

**Figure 1.6** From the lunar orbit of *Apollo 8*, this remarkable photograph captures Earth rising over the nearby Moon.

is 1179 semidiameters of the Earth distant. And the Moon, when she is farthest from the earth, is 68 semidiameters of the earth and 1/3." Dee's distance to the Moon was surprisingly close, at 272,000 miles, just a bit more than the current distance we know of 240,000 miles. But the Sun, which Dee thought was less than 5 million miles away, was way off its currently understood distance of 93,000,000 miles. If no one had been inspired by that event or other happenings in the sky, we would not be much farther along today in our understanding of how the cosmos works.

The pioneering observations of the early modern astronomers, inaccurate as they were, started us down a road that led to our understanding of how far the Sun and the Moon are, and hence the size of the Universe. On December 4, 1639, Jeremiah Horrocks trained his telescope on the late afternoon Sun and saw the silhouette of Venus passing across it. Years later, Edmond Halley would suggest a way to use these rare transits to calculate the distance from Earth to Sun. Learning the true distance between the Earth and the Sun would form the basic yardstick of space.

Times and distances have changed, but what hasn't altered is the wonder that strikes anyone who views a total eclipse. All the knowledge and

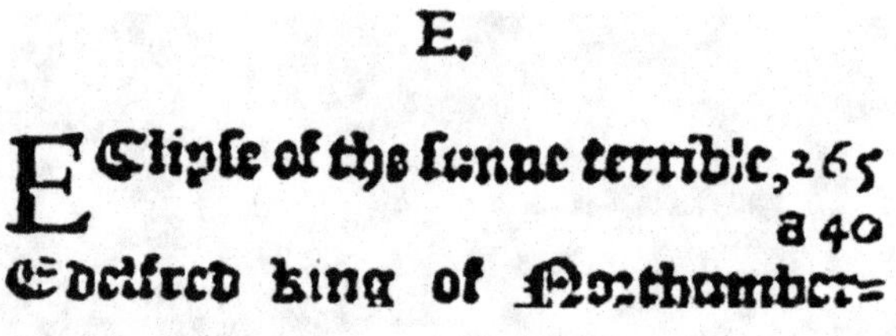

**E.**

E Clipſe of the ſunne terrible,265
                                    8 40
Edclfrcd king of Northumber=

**Figure 1.7** Holinshed's *Chronicles* was an extremely useful resource for several reasons, not the least of which is its thorough index, which shows not only page but also section and line numbers. This image shows the result of a resourceful compositor, who instead of inserting a standard letter "c", spelt the word eclipse with a sketch of a deep partial eclipse of the Sun. Photo by David H. Levy.

understanding that we have gained over the centuries has left that wonder undiminished. It is my hope that, through this book, you will also become inspired to view eclipses, and in turn to become fascinated with the sky that produces them.

ENDNOTES

1  James Joyce, *Ulysses*. New York: Penguin edition, 1968, 210.

2  *King Lear*, I.2.101.

3  *King Lear*, I.2.115.

4  G. P. Akrigg, *Letters of King James VI and I.* Berkeley: University of California Press, 1984, 264–266.

## 2

# Einstein, relativity, and the solar eclipse of 1919

The sun was setting on the links,
The moon looked down serene,
The caddies all had gone to bed,
But still there could be seen
Two players lingering by the trap
That guards the thirteenth green.
The Einstein and the Eddington
Were counting up their score;
The Einstein's card showed ninety-eight
And Eddington's was more. . . .
The shortest line, Einstein replied,
Is not the one that's straight;
It curves around upon itself
Much like a figure eight,
And if you go too rapidly
You will arrive too late.

(W. H. Williams, "The Einstein and the Eddington," 1919, based upon Lewis Carroll's "The Walrus and the Carpenter," 1872)

A total eclipse of the Sun is one of the most thrilling events that Nature has to offer. That one thought is enough motivation to travel the length and breadth of the world to see one. But, in the past, astronomers learned amazing things about our solar neighborhood from watching one. In the second millennium BCE, according to the *Shu Ching*, the Chinese observed eclipses and became so proficient in predicting them that two of their astronomers, Hsi and Ho, were put to death after they failed to predict one. This story is apocryphal. It is unlikely that astronomers were really able to predict eclipses accurately that

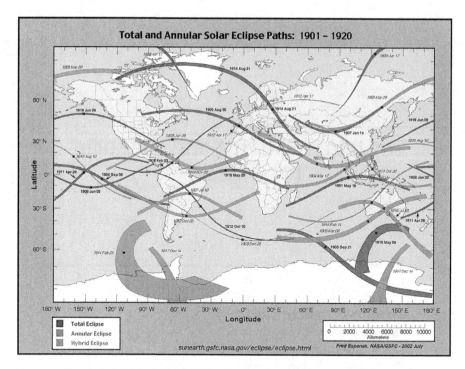

**Figure 2.1** The eclipse of 1919. This important map, prepared by Fred Espenak and his team at NASA's Goddard Space Flight Center, shows the paths of total and annular eclipses that occurred between 1900 and 1920.The map shows the "Leslie Peltier" eclipse of June 8, 1918, crossing the midsection of the United States, and the May 1919 eclipse that tracked across South America, the Atlantic Ocean, and Africa, and which gave Eddington and his team the chance to produce evidence to support Einstein's general theory of relativity.

long ago. By the sixth century BCE the Chaldeans might have been able to make rough predictions of eclipses.

Perhaps the first eclipse to teach us something about our solar system took place in 129 BCE. Its path of totality passed through Greece, where Hipparchus measured the positions of the Moon's shadow. His resulting estimate was that the Moon was 45 times the Earth's diameter, or about 360,000 miles. More than five centuries later, on July 19, 418, Philostorgius became the first to report a new comet during a total solar eclipse. There are many other examples, but none so odd, so important, and so apparently unrelated as the one that took place 1500 years later.

In 1919 Arthur Eddington used a total eclipse of the Sun to redefine our understanding of the universe. This was not a development that increased our understanding of the Sun and Moon, or eclipses. It was, however, an advance that changed completely our understanding of the meaning of gravity.

## How an eclipse led to a new definition of gravity

Picture a beam of light from a distant star traveling through space. For years it moves along a straight line, past interstellar dust and through the vastness of empty space. In the last second it hurtles through the atmosphere and finally stops as it enters your eye. Its journey is a simple one – unless, while on its way to us, it happens to pass a large massive object like a star. For example, in 1919 a group of astronomers photographed the Hyades, when the cluster was placed within a couple of degrees of the Sun in the sky. The stars on the northwest side of the Hyades were closer to the position of the Sun than the bright star Aldebaran, and their light was bent by the gravity of the Sun, giving them slightly different positions in the sky.

What does this bending of light have to do with eclipses? In 1915, Albert Einstein's general theory of relativity offered a new definition of gravitation that related it to space and time. Newton introduced us to gravity as a force. Einstein's theory describes gravity as geometry. Any object moving in space follows a geometric path shaped by the unified effect of mass and energy. The practical result of both Newton's and Einstein's theories is generally the same. Newton's laws will get us to the Moon and home again. But when a beam of energy approaches a massive object, like the Sun, Newton's laws fail.

This theory was fascinating, but as long as it could not be tested in practice, its value would remain unknown. In 1918, the British astronomer Arthur Eddington noted that Einstein's theory could be tested "with regard to the deflection of light by a gravitational field," and that it could be tested by photographing a star near the Sun, and then comparing its position with that on other photographs taken when the star is far from the Sun.[1] The only time that a star close to the Sun would be visible and measurable, of course, would be during a total eclipse. If Einstein were correct, the star would appear deflected away from the Sun, by the same amount as the total deflection by gravity.

Imagine Eddington's pleasure when he discovered that during the eclipse of May 1919, the Sun would be very close to the largest grouping of moderately bright stars in the sky – the open star cluster called the Hyades.

**Figure 2.2** David stands next to Einstein's telescope, displayed now at the Hebrew University of Jerusalem. The telescope is behind him. A 10-inch reflector is at far right, and Minerva, the 6-inch reflector with which he observed the 2006 total eclipse, is in front. Photograph by Wendee Levy.

"On May 29 [the Sun, in eclipse,] is in the midst of a quite exceptional patch of bright stars – part of the Hyades – by far the best star-field encountered. Now if this problem had been put forward at some other period of history, it might have been necessary to wait some thousands of years for a total eclipse of the sun to happen on the lucky date."[2] Eddington developed a plan to photograph stars in the Hyades star cluster, from Principe, a small island off Africa's west coast, which on May 29, 1919, would be under the darkness of a total eclipse of the Sun. The cluster would be just south of the eclipsed Sun, its stars bright enough to be captured on the photographic films of the time.

Five months before the expedition began, Eddington photographed the Hyades field from England using the same telescope as would be brought to Africa. With the Hyades far from the Sun so long before the eclipse, this photograph would serve as a base for comparison.

Eddington arrived in Principe with a 13-inch diameter, 11-foot, 4-inch long refractor. They stopped the lens down to 8 inches to improve the

**Figure 2.3** Low tide at the Hantsport wharf on Nova Scotia's Minas Basin.
The photographer walked out onto a special spot to create this photograph.
Photograph by Roy Bishop.

sharpness of its images. The telescope was not set up in the usual way, on a heavy equatorial mounting. To save weight, it was mounted in a fixed position, and a moveable mirror, or coelostat, directed the light from stars into the lens of the telescope.

Eclipse day dawned with heavy rain. "The rain stopped about noon and about 1:30 when the partial phase was well advanced, we began to get a glimpse of the sun," Eddington wrote. "We had to carry out our programme of photographs in faith. I did not see the eclipse, being too busy changing plates, except for one glance to make sure it had begun and another half-way through to see how much cloud there was. We took 16 photographs. They are all good of the sun, showing a very remarkable prominence; but the

**Figure 2.4** The same shot, taken 6 hours later at an unusually high astronomical tide. To arrive at the same position, Roy had to walk out blindly and very carefully; he could not have his automobile in this second picture. The tide was amplified by the correct position of the Moon and by a strong low-pressure weather system moving through the area. Photograph by Roy Bishop.

cloud has interfered with the star images. The last six photographs show a few images [of starts in the Hyades cluster] which I hope will give us what we need. . . ."[3]

Eddington's 16 exposures ranged in time from 2 to 20 seconds. The first ones did not record any stars, but they did capture the prominence. By sheer luck, helped along with the continuing drop in temperature, the cloud dissipated in the last minutes of total eclipse, and one photographic plate recorded five stars.

Eddington made his first measurements at the eclipse site a few days after the event. The two photos were placed "film to film" in the measuring machine so that the star images were close to identical: "June 3. We developed the photographs, 2 each night for 6 nights after the eclipse, and I spent the whole day measuring. The cloudy weather upset my plans and I had to treat the measures in a different way from what I intended, consequently I have not been able to make any preliminary announcement of the result. But the one plate that I measured gave a result agreeing with Einstein."[4]

For a result as crucial as this one, Eddington had to make sure that his calculations took into account any errors caused by the telescope or the photography. Eddington photographed a different star field, at night from Africa. The same field was photographed also from England. If the "Einstein deflection" were the result of a telescope error of some kind, it would have turned up in these check plates. Once their extensive journey back to England was over, four more plates were processed and examined. One of them confirmed the result shown on the first successful photograph.

Not wishing to take any chances with weather or telescope problems, the expedition also chose a second site much further west along the path of totality, in the South American country of Brazil. There the weather was completely clear, but hot. The team remained at their site for two additional months in order to photograph the same region of sky under cover of morning darkness. They had two telescopes, one similar to the African scope, and a much longer, 19-foot long, 4-inch diameter refractor.

The Brazil expedition measured the positions of the stars on their plates. They did not agree with Einstein's prediction, but Eddington did not panic. He suspected that the cause was deflection not by gravity but by instrument: failure under the clear Brazil sky could be attributed to the heat distorting the coelostat near the telescope's objective lens; in this one case, bad weather at the primary site helped: "at Principe," he wrote, "there could be no evil effects from the sun's rays on the mirror, for the sun had withdrawn all too shyly behind the veil of cloud."[5]

The final verdict on relativity thus had to wait until a measuring engine, a device on which photographic plates are inserted so that the star positions could be measured, was modified to accept the seven oddly-sized plates taken through their 4-inch refractor from Brazil. These plates seemed ideal, their images were perfect, and the result confirmed Einstein's theory.

"Lieber Herr Eddington!" Einstein wrote in great excitement. "I should like to congratulate you on the success of this difficult expedition. I am amazed at the interest which my English colleagues have taken in the theory in spite of its difficulty . . . If it were proved that this effect does not exist in nature, then the whole theory would have to be abandoned!".[6]

One eclipse, one theory. Next time you watch an eclipse of the Sun, share the excitement that Eddington and Einstein felt all those years ago.

### ENDNOTES

1  A. Vibert Douglas, *The Life of Arthur Stanley Eddington*. London: Thomas Nelson and Sons, 1956, 39.

2  Sir Arthur Eddington, *Space, Time, and Gravitation: An Outline of the General Relativity Theory*. Cambridge: Cambridge University Press, 1920, 113.

3  Douglas, 39.

4  Douglas, 39.

5  Eddington, 117.

6  Douglas, 40–41.

# 3

# What causes solar and lunar eclipses?

By the clock, 'tis day,
And yet dark night strangles the traveling lamp:
Is't night's predominance, or the day's shame,
That darkness does the face of earth entomb,
When living light should kiss it?

(Shakespeare, *Macbeth*, 2.4.6–10)

I love asking children what happens when the Moon gets in between the Sun and the Earth. The answer, of course, is a solar eclipse. And what happens when the Earth gets in between the Moon and the Sun. As the Earth's shadow strikes the Moon, we have a lunar eclipse. So what, I finally ask, happens when the Sun gets between the Earth and the Moon? I love their expressions as they start to think of the absurdity of that ever happening. It is a question that makes them think about the wonderful dance of worlds that results in these eclipses. The most important thing to remember about eclipses, then, is not their cause, but their beauty. For lunar eclipses, our thoughts can ruminate on how the shadow of our troubled Earth can move so gracefully across the Moon, as Thomas Hardy noted after the eclipse of 1903, and whose poem is printed at the opening of the chapter on total lunar eclipses (p. 112).

I think that this is the right way to think of a lunar eclipse, because the shadow that falls upon the Moon is that of the Earth – all the Earth – from its continents and oceans to its people and nations. We can take the opportunity to note that the shadow is not full of multicolored nations, but a single shadow. The only difference from eclipse to eclipse has to do not with war and disease but with the unpredictable change of light caused by volcanic

**Figure 3.1** A photograph that appeared in the now-defunct *Montreal Star* after the October 2, 1959 eclipse. It shows the closing partial phase of the eclipse just after the Sun appeared for the first time that day after clouds.

eruptions. If a volcano erupts in the months preceding a lunar eclipse, the dust it sends into the Earth's atmosphere makes the shadow darker. On December 30, 1963, the shadow was unbelievably dense. The eclipse also took place during one of the coldest nights I had ever seen, but we observed from the second storey of my parents' home. As the partial phases began in the predawn hours of December 30, 1963, the Moon seemed to vanish. At totality the Moon simply disappeared.

It might have been the darkest lunar eclipse since 1601. According to *Sky and Telescope*'s report in its March 1964 issue, Walter Haas, the founder of the Association of Lunar and Planetary Observers, who was observing from his home in Las Cruces, New Mexico, claimed that "the Moon sometimes, to the eye, disappears completely in eclipse!" Some observers even reported the Moon disappearing as seen through telescopes. I remember just that blank part of sky where the Moon must have been.

Why was this eclipse so dark? In 1963, the Gunung Agung volcano erupted in Bali and showered dusty material high into the upper atmosphere. The reason that particular shadow was so black was that the material, as it surrounded the Earth, darkened the shadow of our planet. I think it is interesting that with everything we hear on the news about our planet, political and otherwise, none of these issues stretch into space along the shadow. Only some dust and dirt from volcano eruptions – no war, pestilence, or political decision has that ability to affect the shadow. This is what is important about eclipses; they tell us something important about our planet, its atmosphere, and ourselves.

### Why do we have eclipses?

The Moon circles the Earth once every 29.5 days, a period loosely coinciding with a month; the word "month," in fact, is derived from the German word for "moon." Twice each orbit, the Moon, Sun, and Earth line up approximately. The Earth orbits the Sun in a near-circle once every year, and the Moon orbits the Earth, reaching the Sun's position, every 29.5 days. However, the orbit of the Moon about the Earth and the Earth's orbit around the Sun are tilted by a small amount – about 5 degrees – relative to each other; if they were not, eclipses would happen every two weeks: a solar eclipse every new moon and a lunar eclipse at every full moon. Instead, the Moon spends part of its month-long orbit below the plane of the Earth's orbit around the Sun, and part of it above that plane. Twice each month the Moon crosses the plane of Earth's orbit. The two points of crossing, or intersection, of the two orbits are called nodes.

If the Moon is moving northwards in its orbit, it's called the ascending node; if it is going south, it crosses the descending node. The node crossings take place at different phases of the Moon each month. If these crossings take place near new or full moon, an eclipse can occur.

### The saros

In Chapter 1, we saw how the people of England saw an eclipse of the Sun in October 1605. Many readers of this book may have seen the same eclipse, for eclipses repeat themselves every 18 years, 11 days, 8 hours (or 6585.3 days). The eclipse over London was part of a cycle, or *saros*, called saros

**Figure 3.2** My first total eclipse was on July 20, 1963. My view was very much like that in this photograph, taken nearby from a site northwest of mine. The weather was also mostly cloudy, which it was at my site that afternoon. We were pretty certain we were going to miss totality but a slight clearing allowed us to see the corona. In the picture the corona does appear slightly oval, even though the solar cycle was approaching minimum at the time. Photograph by Constantine Papacosmas and used with his kind permission.

137. That same saros happened to produce a deep partial eclipse over much of the southern United States in 2002. In a real sense, everyone who saw the Moon darken the Sun that year also experienced the ancient eclipse of 1605 half a world away.

My first total eclipse was July 20, 1963. It was part of saros 145. Eighteen years and a third of a month later, the Moon's shadow touched down over the Soviet Union. I missed the eclipse that time. But another 18 summers later, the shadow would sweep out of the sky once more, touching the Earth in the Atlantic Ocean some 350 kilometers southeast of Halifax, Nova Scotia, on August 11, 1999. In August 2017 the eclipse will bathe the Earth once more, this time over the United States.

The Chaldean civilization, so successful in monitoring the apparent motions of the Moon and Sun, the planets and the stars, discovered this wonderful, simple saros concept. The saros happens because it takes advantage of three types of lunar cycle. The first is the regular *synodic month*, the time from new moon to new moon. This period is 29 days, 12 hours, and 44 minutes. There is a second type of month called a *draconic month* which is the time from one node to the next similar node. That period is two days less, 27 days, 5 hours, 6 minutes. Finally there is the *anomalistic month*, the time when the Moon goes from its perigee (closest point to the Earth) to the next one. That is 27 days, 13 hours, and 19 minutes. These periods line up about every 18 years.

So eclipses repeat. But because of that critical third of a day, the next eclipse will be a third of the way around the world. For an eclipse to be almost in the same place on Earth, we wait three saros cycles, or an *exiligmos*. The 2017 eclipse of the Sun will be an exiligmos of the eclipse of July 20, 1963, and will follow a very similar path, just farther south.

## The Metonic cycle: a different kind of sequence

Since 430 BCE, when the Greek astronomer Meton discovered it, we have known that the sequence of phases of the Moon are the same every 19 years. The Metonic cycle is a 19-year cycle covering 235 lunar months. Phases of the Moon during this cycle land on the same days during the year, and the cycle was crucial in trying to build a calendar. Often when an eclipse occurs on the day of a particular phase in that cycle, another one will take place at the next cycle. There was a new moon, and an eclipse, on February 26, 1979, one I saw in full glory near Winnipeg, Manitoba. When I began planning to see the February 26, 1998 eclipse in the Caribbean, I recalled the earlier one. The new moon on October 2, 1959 – my first eclipse – was followed by another new moon, and a partial eclipse in the eastern hemisphere, on October 2, 1978, and I saw an annular eclipse of the Sun from Madrid on October 3, 2005 (which was really still October 2 over North America). However, this cycle is not quite as regular as the saros, since it skips one of every five repetitions. So there was not an eclipse on October 2, 1997. Moreover, unlike the saros, the Metonic cycle can't be used to predict where an eclipse will take place.

There is yet a third kind of cycle, called the *Inex cycle*. It features solar eclipses that occur at approximately the same place on Earth but at opposite latitudes on Earth. An inex lasts 258 lunations, or synodic months. Although the famous comet astronomer A.C.D. Crommelin first suggested the idea of a different kind of cycle in 1901, George van den Bergh studied and named it during the 1950s.

From the unpredictable eruption of a volcano that darkens our atmosphere, to the completely predictable dancing of the shadows of Earth and Moon, eclipses are captivating events that can inspire and teach us about our solar neighborhood, its aspects and motions, and our history.

Some terms to review:

The *ecliptic* is the Sun's apparent path through the sky each year, as seen from Earth. The Moon's *ascending node* is the place in the Moon's orbit around the Earth where it crosses, moving north, the orbit of the Earth as it travels round the Sun. The Moon's *descending node* is the place in the Moon's orbit where it crosses, moving south, the orbit of the Earth. An *eclipse season* is a short interval of time, lasting about a month, in which the Sun's apparent motion along the ecliptic places it close to one of the Moon's nodes. Finally, an *eclipse year* is the time for the Sun's apparent motion to carry it from one ascending node of the Moon back to the ascending node. Eclipse years last 346.6 days.

**Figure 3.3** Clouded out, but beautiful. A scene of beautiful desolation on Nova Scotia's south shore, at Crystal Beach. There we attempted to observe the 1970 eclipse but we were clouded out. What we did see was a darkened coastline and island out to sea, along with the edge of the Moon's shadow. It was nevertheless a thrilling experience.

## The eclipse members of saros 126

It is fun to trace the eclipses of a particular saros. I choose solar saros 126, because that cycle included both the 2008 Russia eclipse and the "Leslie Peltier eclipse" of 1918. This saros was born with a tiny, barely detectable partial eclipse near the South Pole, on March 10, 1179. Successive partial eclipses continued, each one getting a little larger in scope and also heading further north, through 1197 and the thirteenth and fourteenth centuries. On June 14, 1341, the first annular eclipse took place. These annular eclipses tended to run for about five minutes each until the final five or six, during which annularity shortenened until it was just a few seconds long. Then, on April 14, 1828, the first hybrid eclipse occurred. It was annular at the sunrise and sunset points on the track, but because of the slight decrease in distance between Moon and Earth at midday, it was total briefly at greatest eclipse. The saros produced its first total eclipse on May 17, 1882. We know that the track of the June 8, 1918 eclipse passed over the central hub of the United States, and that the track hit the U.S. again on June 30, 1954. On July 10, 1972,

Now circling in peace...

**Figure 3.4** Phases of the Moon. This single photograph is a multiple exposure of four pictures of the Moon, each at a different phase. The exposures were taken at the Adirondack Astronomy Retreat in August, 2007 as the Moon rose over nearby mountains each night. Photographs by David H. Levy.

thousands of people saw the "ALCAN" eclipse that crossed over Alaska and Canada. On July 22, 1990, Steve Edberg and I caught the opening seconds of the partial at sunset, and Wendee and I enjoyed the total eclipse of 2008 as it tracked over Russia. On August 12, 2026, the shadow will pass over Greenland, and near the North Pole on August 23, 2044. Thereafter follows a series of partials that end with the eclipse of May 3, 2459 near the North Pole.

Could any of Shakespeare's countrymen have seen the eclipses of this saros? Only if they could travel: One took place off the coast of Africa, another concentrated on the Indian Ocean. None crossed over England.

# Part II OBSERVING SOLAR ECLIPSES

# 4

## Safety considerations during a solar eclipse

> Clouds and eclipses stain both moon and sun . . .
>
> (Shakespeare, *Sonnet 35*.3)

Any solar eclipse – even the tiniest of partial eclipses – is an event worth watching. I believe that Shakespeare and his colleagues, in a little studied, probably multi-authored play *Edward III*, were describing a total eclipse even though this was years before the great solar eclipse of 1605. (The full quote opens Chapter 5.) Although it is highly unlikely that Shakespeare ever witnessed a total eclipse of the Sun, he probably heard or read reports about them from those who had. Also, he had the opportunity to witness several partial eclipses from London, including one on July 31, 1590, during which half the Sun was covered by the Moon at maximum. There were other, shallower ones, like the sunset eclipse on August 11, 1589. More important, the great writer's fertile imagination would have allowed him to imagine what a solar eclipse, continued to totality, might have looked like. It is this sense of imagination and dreaming that leads me to believe that he had an eclipse in mind when he wrote (if he wrote) the lines that begin the next chapter; this was a play of multiple authorship.

In July 1990 Steve Edberg and I drove for several hours along back roads to a spot in Northern California where we watched the Sun begin to set. It looked like a normal Sun until it was about half way down, and the first nick of the Moon announced the beginning of a solar eclipse that was total far north and west of us, in Finland. In north central California, the Sun set about a minute after the eclipse began. We may have been witness to the last particular event of this eclipse, the point on Earth at which the eclipse began at the moment of sunset.

**Figure 4.1** Moonset? On a clear evening at our home in Vail, Arizona, I decided to take a brief exposure of the Moon, with Venus just a bit to its west and south. Just after the exposure began, Wendee reminded me that our favorite television program, *The West Wing*, was about to begin. When the program ended I rushed out and stopped the exposure. The resulting image shows the Moon as a long bright streak setting in the west, Venus setting next to it, and the tracks of several airplanes descending to land at Tucson International Airport. Photograph by David H. Levy.

Total eclipses are rare and magnificent, but partials are much more frequent. Because they also show the motion of the solar system in real time, they can be inspiring events. If one occurs near you, be sure not to miss the show.

I am sure that you've all heard of the dangers involved in observing a solar eclipse. Children have been shut up in their classrooms with blinds drawn to avoid the peril to their eyes during a total eclipse. Let's try to replace these unnecessary and shortsighted actions with facts:

(1) The Sun is *always* dangerous to look at with unprotected eyes, whether there is an eclipse or not. However, when there is no eclipse, the Sun is so bright that your eyes cannot stand the stress of looking at it, and you instinctively turn your head away.

(2) During an eclipse, as the Moon covers more and more of the Sun, two things happen. The Sun's crescent shape and low brightness make it

**Figure 4.2** An owl looks at the Moon. A beautiful great horned owl poses atop a piece of vertical pipe stationed near my observatory. The owl stayed there patiently while I composed the picture, with owl and gibbous Moon placed next to each other. Photograph by David H. Levy.

easier and more interesting than usual to look at. But even though the Sun is dimmed by the Moon, the visible part of the Sun, per unit area, is as bright and dangerous as ever. As you look, your built-in urge to turn away isn't there just at the time that looking at the Sun is most dangerous. There have been documented cases of people losing their eyesight by prolonged staring at the Sun even when there was no eclipse. One well known late 1960s news item concerned a story about a group of youngsters on LSD who stared at the Sun for some time, perhaps several hours, and were permanently blinded. During the 1979 eclipse, at a Las Cruces school a fifth-grader was temporarily blinded for about three days after staring too long at the Sun. Eclipse blindness can last hours, days, months, or a lifetime, and so I endorse

Nanette's River, November 19, 2005
Photo by David and Mark

**Figure 4.3** The magic of eclipses, both of the Sun and Moon, is limited, unfortunately, to those rare moments when eclipses actually occur. But the majesty of the sky is there every night for all to enjoy. An interesting chain of stars appears in Caasiopeia. I located this ethereal grouping just before dawn, while comet hunting on May 3, 2001. In this photograph, exposed using Obadiah, my 12-inch Schmidt camera, the stars that are part of the grouping are artificially "enhanced" using Adobe Photoshop. Since the grouping, or asterism, resembles a flowing brook, I informally dubbed it "Nanette's River" in honor of our daughter. Photograph by Mark Vigil and David H. Levy.

the standard recommendation: *Never look at the Sun at anytime, but especially during an eclipse, without proper eye protection.* Despite the many published warnings before the great European eclipse of 1999, ophthamologists in Europe recorded many eye injuries. Their examinations detected images of crescent suns, of different thicknesses representing different stages of the eclipse, burnt into each retina.

The *only* time that it is completely safe to look at a solar eclipse is during its total phase, when 100% of the photosphere is covered by the Moon. But this magnificent phase is surrounded by the two most dangerous parts – the minute or so immediately before and after totality. During these times, it is possible to see the dark Moon even though the Sun is still shining brightly, and its thin line of sunlight is highly dangerous to your retinas.

When it comes to expertise on eye safety during solar eclipses, the person I trust more than anyone else is Dr. B. Ralph Chou, associate professor at Canada's University of Waterloo School of Optometry. He has spent a lifetime chasing and observing solar eclipses, and studying the effects of the unfiltered Sun on the human retina. The real issue, he notes, is a condition called retinal burns. "Exposure of the retina to intense visible light causes damage to its light-sensitive rod and cone cells," he writes. "The light triggers a series of complex chemical reactions within the cells which damages their ability to respond to a visual stimulus, and in extreme cases, can destroy them. The result is a loss of visual function which may be either temporary or permanent, depending on the severity of the damage. When a person looks repeatedly or for a long time at the Sun without proper protection for the eyes, this photochemical retinal damage may be accompanied by a thermal injury – the high level of visible and near-infrared radiation causes heating that literally cooks the exposed tissue. This thermal injury or photocoagulation destroys the rods and cones, creating a small blind area."[1]

One of the safest means of observing an eclipse is by projection. With the unaided eye, a pinhole projector can work. Insert a hole into a piece of cardboard – a very small pinhole – and the place the cardboard behind you so that you can project the pinhole images on the ground, table, or a second piece of white cardboard. Aside from the image being pretty small, this is an effective means. However, *do not look through the pinhole at the Sun*. You can also use a small telescope to *project* an image of the Sun onto a piece of white cardboard. If you use a telescope, use a Newtonian reflector or refractor. A compound optical system like a Schmidt–Cassegrain has internal plastic parts that can catch fire if the telescope is left pointing at the Sun.

One of the safest traditional viewing glasses for eclipses is number 14 Welders glasses. These can be purchased from welder's supply stores. Do not use lower number glasses. Even more popular are sheets of aluminized mylar – but use only the ones made specifically to observe the Sun – these are sold, or given away, at planetaria and science stores when an eclipse is about to occur.

The following items are *not* safe:

> black and white exposed film containing images
> medical X-rays
> color film
> smoked glass
> multiple pairs of sunglasses
> mylar blankets designed for warmth on cold nights
> mylar used in gardening
> CDs or DVDs

neutral density filters

polarizing filters.

None of these come close to passing Chou's optical tests: "A safe solar filter," he records, "should transmit less than 0.003% (density ~4.5) of visible light (380–780 nm) and no more than 0.5% (density~2.3) of the near-infrared radiation (780–1400 nm)."[2]

I have always felt unhappy by stories of children locked in classrooms to avoid seeing an eclipse of the Sun. The procedure is just plain wrong on several levels. Children are being deprived of one of Nature's grandest spectacles, and they are deprived of a rare opportunity of witnessing the solar system in motion, in real time. Moreover, a few hours later when they learn on the evening news that thousands of people viewed the eclipse safely with easily obtainable equipment, they will realize that their teachers have been overly cautious with them. An eclipse is a major natural event that is worth the trouble to observe safely.

ENDNOTES

[1] B. Ralph Chou, Eye safety during solar eclipses, *NASA RP 1383 Total Solar Eclipse of 1999 August 11*, April 1997, p. 19.

[2] *Ibid.*

# 5

## What to expect during a partial eclipse of the Sun

King of France: A sudden darkness hath defaced the sky,
The winds are crept into their caves for fear,
The leaves move not, the world is hushed and still,
The birds cease singing and the wand'ring brooks
Murmer no wonted greeting to their shores.
Silence attends some wonder and expecteth
That heaven should pronounce some prophecy.
Where or from whom proceeds this silence, Charles?

Dauphin: Our men with open mouths and staring eyes
Look on each other as they did attend
Each other's words, and yet no creature speaks,
A tongue-tied fear hath made a midnight hour,
And speeches sleep through all the waking regions.

King of France: But now the pompous sun in all his pride
Looked through his golden coach upon the world,
And, on a sudden, hath he hid himself,
That now the under earth is as a grave,
Dark, deadly, silent and uncomfortable.

(Shakespeare *et al.*, *Edward III*, 13.1–18)

### The eclipse of October 2, 1959

When I go back into the earliest of my observing records, I find that session number 1 – my very first recorded observing session – took place on October 2, 1959. That year a full total eclipse tracked by the coast of Massachusetts, and in Montreal the Sun would rise in a deep partial eclipse. It was Rosh HaShana, our Jewish New Year, and we were missing school for it, so Mom was happy to drive my brother Gerry and me to a site with a good viewing location to the east. At the Mount Royal Lookout we saw nothing but clouds but Mom noted clearing toward the west. We piled back into the car and headed toward Westmount Lookout. Finally the clouds broke off a little, enough to let us see the closing portion of the eclipse. It seemed to be nothing – just the Sun with a small round nick in it – but to me it was everything. I decided to keep a record of this observing session, and hence all the observing sessions I'd have. As I write these words, this morning I completed session number 15455M2, held a few hours ago under another predawn partly clouded sky.

This eclipse had a profound effect on me. For the first time I sensed that the heavens were dynamic, changing. We actually watched more and more of the Sun reveal itself as the Moon continued its retreat. In real time we saw the Moon orbiting the Earth! Eclipses prove that the sky changes. The 1959 eclipse inspired me to read more about these wonderful eclipses. Using Herbert Zim and Robert Baker's hard-to-beat Golden Nature Guide *Stars*, I learned that eclipses take place when the Moon passes a particular place in its orbit around the Earth, a place called a *node*. The book explained how eclipses work, but more than anything else it impressed upon me the necessity to start watching for these events. The eclipse of 1959 was but an *hors d'oeuvre* or appetizer to whet the palate for larger and better eclipses to come. I learned from that book, for example, that a total eclipse would tear across eastern Canada in just four years, on July 20, 1963. I resolved then to try to see this eclipse.

Like the one in this book, that book contained a chart of eclipses. When an eclipse is total it carves a path of darkness along the curved surface of the Earth. Somehow my concentration was drawn to a particularly short line, way up in the North Atlantic Ocean between Greenland and Iceland. On that day, I learned that there would be a very brief full eclipse, with a flash of totality lasting about two seconds.

I never forgot reading about that eclipse, even until the actual dawn of February 3, 1986. On that morning I drove quickly the 100 miles from my home southeast of Tucson to the ghost town of Steins, New Mexico, because I knew that on that day a brief total eclipse would be taking place near Iceland. Whatever do freezing Iceland and sunbaked Steins have in common? Just that both would be under the shadow of the Moon that day, only Steins would get just a nick of the outermost part of the penumbral part of the Moon's shadow for about 20 minutes. Steins was just at the edge of the

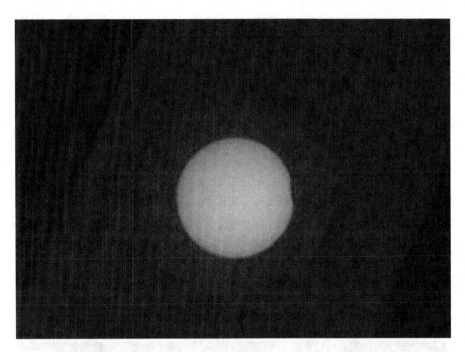

**Figure 5.1** The opening partial phase of the eclipse of February 26, 1979 is pictured here. Photograph by David H. Levy.

eclipse's penumbral phase. Excited to get there, I drove too quickly and got a speeding ticket as I rushed through the town of Benson, but undaunted I drove on, arriving at Steins about a half hour before first contact. Right on schedule, a small nick developed in the Sun, spread into a paper cut, and then disappeared. On my way home I stopped in Benson, paid the ticket, and headed home.

Although the tiny eclipse of 1986 helped to satisfy my curiosity about the nature of these beautiful events, it also whetted my appetite for more. Eighteen years later the eclipse took place again, and this time the shadow fell quite a bit westward. No longer would this saros cycle produce total, hybrid, or annular eclipses. The remaining eclipses would all be partial, including the eclipse of October 14, 2004 that could be seen over the Pacific. Not wanting to travel too far, I contacted Gary Fujihara of the University of Hawaii's Institute for Astronomy, who kindly invited me to come to Hilo, on the "Big Island" of Hawaii, to lecture about astronomy. While there I'd also have a chance to ascend to the summit of Mauna Kea to observe the eclipse with Steve O'Meara, one of my oldest friends, and a group of astronomers assigned to Japan's Subaru telescope.

On eclipse day, we drove to the summit of the tall volcano. I was happy that the volcano was quiescent, but always I was just a little concerned that it is not considered extinct. I recall when the first director of the Canada–France–Hawaii Telescope, Riccardo Giacconi (who would later win the 2002 Nobel Prize in Physics for his studies of cosmic X-ray sources)[1] lectured about the then-new Canada–France–Hawaii Telescope at a meeting of the Royal Astronomical Society of Canada in June 1980. While noting that the mountain was not expected to erupt for a long time into the future, he did display a cartoon showing the summit blowing its top, launching CFHT, dome, scope, and mount in place, into orbit as a space telescope.

On that afternoon, our caravan arrived at the mid-level facility, built at the 9,000 foot level, and then continued on to the summit. The clouds were still fairly dense there, although the Sun peeked through from time to time. The eclipse was the most dramatic partial I've ever seen. From the almost heavenly

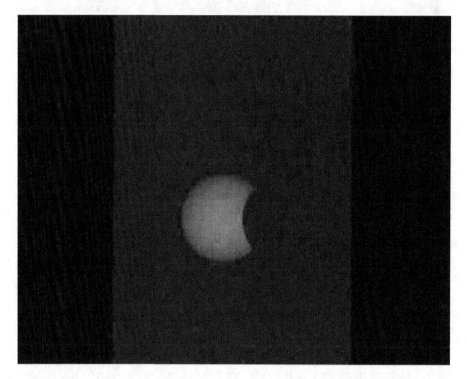

**Figure 5.2** The same eclipse in a deeper partial phase. Photograph by David H. Levy.

altitude of 13,500 feet, the sky was incredibly steady. There was one small spot that before sunset appeared greenish on one side and reddish on the other, as if it were preparing to perform its own green flash before setting. This time, the Moon's umbral shadow passed north of the Earth entirely, so there was no total phase to this eclipse anywhere on Earth.

This saros series that includes the 1986 and 2004 eclipses has a long and fascinating lineage. It produced a total eclipse in 1572, the year of Tycho's great supernova in Cassiopeia (but that eclipse was not visible from England), and another, visible from Africa and India, that accompanied Halley's Comet in 1680. On May 25, 1770, a total eclipse from this series noted the American preparations for independence, which occurred six years later, although the path did not cross anywhere close to North America. In the future, the series will continue to produce smaller and smaller partial eclipses until it ends with a small partial eclipse near the North Pole, on January 9, 2149.

Although the 2004 version of this eclipse is an example of a sunset partial eclipse, it did not provide one of the events I look forward to at such events – a double green flash. As the two contact points between Moon and Sun touch the horizon, each could produce a green flash. Unfortunately, the Sun on that October day set into a deck of clouds, depriving us of that opportunity.

### Activities for a partial eclipse

First contact: timing.

Partial eclipses of the Sun are interesting for a number of reasons. Timing is one of them. If your accurately printed information says that the Moon will take its first bite out of the Sun at precisely 1:23 pm eastern daylight time, that is precisely when the Moon and the Sun first contact each other. We call that first contact. Well-known observer and writer Stephen James O'Meara notes that first contact is not a single event but three discrete events: one when the first touch of the Moon and Sun is seen through a telescope, a second when seen (seconds later) through binoculars, and finally when it is detected with the unaided eye.

Eclipse contacts are definitely events by which you can set your watch. First contact marks the important difference between the Sun appearing whole, and the Sun appearing not quite whole. Within a minute of this contact, a more definite nick will become noticeable.

**Figure 5.3** The 1979 eclipse in a deep partial phase. Photograph by
David H. Levy.

### Sky darkening

Some of the ideas I offer in these pages come from O'Meara's excellent article *An eclipse timetable*. His goal was to encourage observers to get the
most out of traveling to see a total eclipse, but his ideas benefit the observer of
even the smallest of partial eclipses.[2]

If the eclipse is a slight one, where less than half of the Sun is covered,
a casual observer will not notice any significant drop in the ambient light.
The light level can actually be measured qualitatively by recording the color
and quality of the sky, and of any clouds which might be passing by. Record
your results not just on the sky or clouds, but also on nearby objects and
distant landforms. Again, these changes should become noticeable only if
the eclipse involves 50% or more of the Sun. I have devised a qualitative
brightness scale that might be helpful:

10:    Sky and clouds in full daylight. Shadows normal.
 9:    Just the tiniest tinge of fading is noticed. Shadows still normal.

These are the effects that might be noticed during a minor partial eclipse, when less than half the Sun is obscured by the Moon at the maximum point. The key to seeing such an eclipse is just to see it. If the solar activity cycle is such that sunspots are plentiful, you can watch the edge of the Moon obscure sunspots, and later on, release them as it withdraws. Although these events do not advance science, they are exciting events to enjoy. They demonstrate the following:

(a)    that the Moon is orbiting the Earth;
(b)    that the Moon is closer to Earth than the Sun is;
(c)    that in the course of its orbit, the Moon occasionally passes in front of the Sun, causing a solar eclipse.

If the Moon passes in front of a portion of the Sun, the event is called a partial eclipse.

If the Moon passes directly in front of the Sun, we have a central eclipse.

If the Moon happens to be near its closest distance from Earth at the time it crosses the Sun (perigee), we have a total eclipse. These events are rare, happening an average of once every 300 years at any single location. Using fairly simple physics, eclipses can be calculated for thousands of years in the future with great precision.

## Deeper partial eclipses

If a shallow partial eclipse is fun to watch, a deeper one is far more fascinating. If more than half the Sun is covered during the eclipse, the sky will begin to darken noticeably. Shadows will appear sharper. Perhaps least of scientific import, but most important from a cultural point of view, is the notion that the landscape is somehow quieting down. With the rapid decline of light, the landscape takes on the peaceful air of twilight. For me, the eclipse of June 10, 2002 was a good example of a deep partial eclipse. This effect was obvious to me that hot June day, as the temperature decreased and the cycle of nature appeared to slow down.

Since this was only a partial eclipse, albeit a deep one, the darkening effect lasted only a short time. Soon the Moon was "in full retreat, releasing the solar spoils it had won," as Leslie Peltier might have said. The sky soon returned to its previous brightness level, the spots reappeared from their attack by the Moon, and all was as it had started.

In Chapter 1, we explored another eclipse in this same saros. In England, the October 1605 eclipse was a very deep partial, but in southern France and parts of Spain, it was total. Although this particular saros no longer produces total eclipses, it does offer a series of annular eclipses, where, at mid-eclipse the Moon is surrounded by a ring or annulus of sunlight. Most of those who traveled south to the path of annularity were clouded out, but here in Arizona we had a magnificently clear day. As we continue our discussion of the effects of a deep partial eclipse, note how closely they come to the words that Shakespeare and his colleagues chose in *Edward III*, the words chosen for the opening of this chapter. The idea that the wind often drops near the deepest portion of a partial eclipse is mentioned specifically in the quoted passage: "The leaves move not,";[3] moreover, the passage continues with "the world is hushed and still,"[4] which defines a strange stillness typical of a deep partial eclipse. "The birds cease singing"[5] is also a symptom of a deep partial solar eclipse. An eclipse does not have to be total for these effects to be noticed; almost total will suffice. These comparisons and observations are not modern attempts to coincide with old writing; they were just as true then, even more so in an age, prior to the telescope, when observers paid more attention to landscape and seascape.

### For deeper partial eclipses, the brightness scale concluded

For eclipses where more than half the Sun is covered, we now conclude the brightness scale.

8: (Typically after the Sun is more than half covered by the Moon) distant clouds show a slight fading. Nearby objects like street lamp posts are starting to cast sharper shadows. Around the time the Moon covers more than half the Sun, you may notice that, even in mid-summer, you do not need to wear UV protection when looking at the landscape. (However, you must continue to use proper eye protection when looking directly at the Sun.)

7: Clouds and distant objects continue to fade somewhat.

6: Fading more pronounced.

5: Fading continues.

4: The eclipse is getting deeper, and the Sun is starting to appear as a crescent. Look more carefully around you. At this point the surrounding landscape or seascape is starting to resemble twilight, only not so much: the kind of darkness differs from twilight because the Sun is present in the sky.

3: As the Moon covers more than 75% of the Sun's face, over the landscape or seascape, the darkness is starting to accelerate. The rate of darkening is so rapid that it can almost be seen visually.

2: The view is evolving into something unique as shadows continue to darken. As the eclipse deepens, shadows become sharper, including your own. In full sunlight, your shadow is somewhat fuzzy in appearance because it has a small "penumbra" at the edges where just part of the Sun is reaching you. During a deep partial eclipse, your shadow becomes sharper. The side of you facing away from the Sun should be fuzzier, with a more pronounced penumbra, than the side facing toward the Sun.

1: At this point the partial eclipse is very deep and the crescent is very thin. The sky toward the western horizon will be darker than the sky toward the east. The reason is that the umbra of the Moon's shadow is approaching you. If the partial eclipse is this deep, you are probably close to the edge of the path of totality, and a short drive might have gotten you into it. The situation is similar to the great eclipse that occurred over New York on January 24, 1925. People in lower Manhattan and Staten Island saw a very deep partial eclipse of the Sun under a clear and cold sky, with temperatures hovering near zero degrees F (–18 degrees C), bitterly cold. For these people, more than 99% of the Sun was obscured, and they might even had caught a good view of an extended showing of Baily's beads as the Moon slid past the Sun. Observers in the Bronx, parts of Queens, and upper Manhattan were in the Moon's umbral shadow and saw a total eclipse.

0: The eclipse enters the total phase, the subject of Chapter 7. Remaining clouds are darker than they would appear after sunset. No real shadows.

## Partial eclipses and history

Certain partial eclipses have captured as much attention as total eclipses. A famous example is the partial eclipse of December 25, 2000. This "Christmas Day eclipse" featured effects that would be consistent with a slight partial eclipse. I was joined by family and friends to enjoy a mild eclipse during which about a quarter of the Sun was obscured at best. At places further north, the eclipse was deeper; from Montreal about half of the Sun was covered by the Moon at maximum eclipse.

## Closing stages of a partial eclipse

An eclipse isn't over yet after the middle has occurred. It is a good idea to repeat the observations you have made as more of the Sun comes out from the departing Moon.

Solar eclipses are special, partly because they are rare. Even though partial solar eclipses are more common than the lunar type, they occur over a much smaller area and, therefore, from a single location such eclipses take place much less frequently. So long as you have proper protection for your eyes, a solar eclipse is definitely a must-see event. If you already have an interest in astronomy, the eclipse will enhance it. If you just happen to take a couple of hours off work to enjoy an eclipse, the event could inspire you to become interested in other aspects of the sky.

ENDNOTES

1 http://nobelprize.org/nobel_prizes/
  physics/laureates/2002/giacconi-autobio.
  html. Accessed 6 September 2009.
2 S. J. O'Meara, An eclipse timetable, *Sky & Telescope (Eye on the Sky column)*, April 2006, 64.
3 *Edward* III.13.3.
4 *Edward* III.13.3.
5 *Edward* III.13.4.

# 6

# A ring of fire

Crooked eclipses 'gainst his glory fight . . .

(Shakespeare, Sonnet 60.7)

When the Moon, during its monthly orbit of the Earth, *directly crosses* the position of the Sun, the result is a *central eclipse*. If, at the time of its crossing, the Moon is close enough to us that its angular diameter is greater than that of the Sun, a *total eclipse* is the result. These are the eclipses we travel around the world to see, and will be the subject of Chapters 7, 8, and 9. But what if the Moon's angular diameter is *less* than that of the Sun? Then we have what is called an *annular eclipse*, which we can also call a ring eclipse. During these moments the Moon's black silhouette is surrounded by a brilliant ring of sunlight.

As exciting as an annular eclipse might be, it is still a partial eclipse. Thus, all the strict rules about looking directly at the Sun unfortunately apply. And more: because there appears to be so little sunlight, observers are tempted to look directly at the ring of Sun. Thus, if the annularity lasts more than a few seconds, blindness can result. Therefore, use a filter whenever looking at the Sun during all phases of an annular eclipse.

## The annular eclipse of September 2005

Of the several annular eclipses I have seen over the years, none was as inspiring as the October 2, 2005 eclipse in Madrid, Spain. The eclipse might not have featured the gut-wrenching splendor of a total eclipse; it was a quiet wonder as the uneven circle of Bailey's beads began to form, and then turned into a beautifully symmetrical ring. It's not a total eclipse, but an annular

The Moment of Rosh HaShana 5766
Annular Eclipse of the Sun from Madrid, Spain
Photographed by Wendee Levy and presented to Congregation Or Chadash

**Figure 6.1** The October 3, 2005, annular eclipse visible from downtown Madrid. The sky was brilliantly clear, and the photograph was taken by holding the camera next to the telescope eyepiece. Photograph by Wendee Levy.

eclipse of the Sun is a unique experience not to be missed. And in this age of inexpensive hydrogen-alpha filters, on October 3, 2005 we saw the progress of an annular eclipse complete with prominences leaping away from the photosphere.

Traveling to Madrid was not a difficult choice to make. To view an eclipse in the middle of a major historic city that is easy to get to and which is famous for clear and dry weather at this time of year all helped to make the trip attractive. When my wife Wendee and I arrived, we met an assemblage of about 20 amateur astronomers from Switzerland, Sweden, Denmark, and the United States.

As the first rays of the Sun hit Spain's ancient royal palace around 8 am on eclipse morning, the whole city seemed bright with anticipation. It had been a very long time – April 1, 1764 – since Madrid had an annular eclipse, and even then the ring wasn't as perfect as it would be today. When that earlier annular eclipse darkened the sky over Madrid, Charles III was about to move into the brand new Royal Palace, which had just been rebuilt and which is still one of Madrid's grandest structures.

I felt a strong sense of history in Madrid while proceeding to Madrid's Park del Retiro where dozens of telescopes were being set up on the shore of a large pond. One trusting observer had set up several thousand dollars worth of of camera and telescope equipment in this very public place, then just left it alone and unguarded. We couldn't find him among the gathering crowd of observers. The excitement at that park was intense, a further testament to the single language that lovers of the sky have, especially when Nature is about to put on one of her grandest shows.

We drove back to our observing site, on the roof of our hotel. On the way we passed the rail station, the scene of terrorist bombings just 18 months earlier and a reminder that the world of our age can produce both glorious eclipses and tragic human events. Back at the hotel roof, Wendee and I set up Minerva, my 6-inch f/4 reflector. We added a solar filter mounting made out of a flower pot holder, and a hydrogen-alpha filter that would have amazed Charles III. The Moon took its first bite at 9:40 am; although the Sun sported no spots, a magnificent cluster of 13 prominences was on the side opposite to where the Moon had just taken its first bite. With a deepening eclipse the whole city slowly darkened. The temperature plummeted, falling several degrees as a brisk wind picked up, which is the opposite of what usually happens, as the wind normally dies down. A mile or so to the west, the Royal Palace also darkened, just as it did in 1764. Office workers began appearing on rooftops all over the city. As I peered through the eyepiece, the white light filter allowed me to see a thin horseshoe of light. But from her perch at the hydrogen-alpha telescope, Wendee suddenly called out "there is a small red dot appearing at the very open end of the horseshoe!" It was a mighty prominence coming into view. Right at that moment, we heard an announcement from someplace – a passing helicopter or policecar – asking everyone to go indoors to avoid the rays.

Two minutes later, a perfect ring of sunlight, with 13 prominences on one side and one big one on the other, faintly lit the clear sky over Madrid. As the Moon slowly wended its way across the Sun in the following four minutes, through the hydrogen-alpha filter it really looked like a total eclipse, with the darkened center of the Sun surrounded by prominences. The only thing missing was the corona. Looking around from our rooftop perch, we could see the darkened city of Madrid, its pulse stopped by time as people

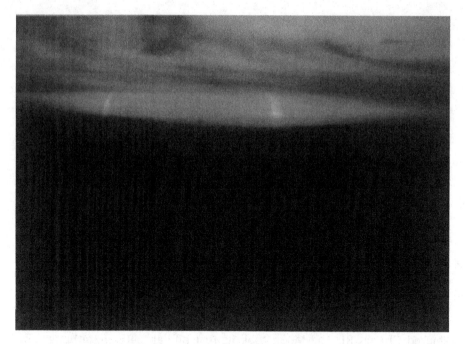

**Figure 6.2** The annular eclipse of January 4, 1992, ended at sunset just on the coast of California. As the afternoon wore on, increasing cloud cover threatened our view of the event. However, the Sun actually set *below* the cloud deck, where, through thinner clouds, we were able to witness the annular phase. Photograph by Tim Hunter.

everywhere stopped what they were doing, donned their eclipse glasses, and looked to the sky.

After annularity, the partial phases took over once again. There were two last contacts, the white light one, however, taking place about 30 seconds before the one seen in hydrogen-alpha, giving us a wonderful view of the Moon exiting the Sun's mighty chromosphere.

## What to see as the annular phase approaches

As the central portion of the eclipse approaches, it advertises itself with an increasingly rapid darkening of the sky. The Sun becomes a thin crescent, which becomes ever more slender as the Moon slowly reaches its key position. However, by now it is obvious that the eclipse will not be total and that the Moon (today, at least) subtends too small an arc to cover the entire Sun. The crescent, though skinny, is very long, curving about until it

**Figure 6.3** As we settled in to observe the eclipse, Gene Shoemaker (fourth from left) suggested he return to Palomar, which was only a few miles away, to do some work. Carolyn's response (she is pictured at right) was that he'd better stay and enjoy the afternoon. "Now that you explain it, I will!" was his response. Thus he became a part of our group picture, which includes friends and colleagues like longtime Palomar observatory manager Robert Thixten at bottom right. Photograph by Carol Hunter.

almost forms a ring. Then, virtually without warning, the following edge of the Moon makes contact with the Sun. This is called second contact, and the annular or "ring" phase has begun.

## Annularity

During the next few minutes the Moon slithers across the face of the Sun. If you have positioned yourself on the centerline of the path, you will, at one point, see the Sun as a perfect ring, its inner and outer edges matching perfectly. Otherwise, the ring will have a thicker and a thinner side.

The longest annular I've ever seen was the event of May 10, 1994 during which the annular phase lasted about nine minutes. The shortest one was on May 30, 1984 from a site just outside of New Orleans. On that day the Sun almost completely covered the Moon, allowing prominences to be visible easily in hydrogen-alpha light, though not in visible light.

### Observing activities recommended for annular eclipses

See Chapter 5 for general advice on observing a partial eclipse. As the eclipse approaches annularity, pay special attention to the following.

Using a safe solar filter only, check to see if the Moon's edge shows any color fringes. With the general level of declining light, similar color fringes may be visible around sunspots that are still visible on the exposed photosphere. These fringes appear when the contrast between the darkening sky and the Moon's limb, or sunspot edge, is magnified by the lower light level present during an eclipse, and are not real but an effect of visual perception. During the annular phase, try repeating this observation for the Moon's edge.

**Figure 6.4** It's hard to see, but for our group it was heaven on Earth as we successfully viewed the setting Sun in annular eclipse from near the summit of Palomar Mountain, California. Photograph by Tim Hunter, used with his kind permission.

Unless the difference in angular size between the Moon and the Sun is very small, even at mid-eclipse the sky will be too bright to spot the corona. However, if the Moon is close enough to the apparent size of the Sun, it might be possible to glimpse the Sun's inner corona during an annular eclipse. During the February 16, 1999 annular eclipse in Western Australia, the famous German amateur astronomer Daniel Fischer successfully observed and photographed both Baily's beads, an effect not often caught during an annular eclipse, as well as the even rarer option of glimpsing the chromosphere and the inner corona. "Then the chromosphere itself was cut into segments by the advancing lunar limb," Fischer writes. "and soon most of it was covered as well. The glaring solar crescent just arc minutes away was not nearly bright enough to disturb the view – and there was still more to come: When the Moon had covered the chromosphere it was obvious that there was something else behind it – the inner corona!"[1]

Our experience in Madrid was just as thrilling, but in a different way. Even before the annular phase began, Wendee noticed prominences appear on the darkened side of the Sun. Astonished, I asked to see as well. Wendee, it turned out, was gazing through the hydrogen-alpha telescope that had been added to the telescope just a few hours earlier. To see prominences in this way during an annular eclipse, not a total eclipse, was extraordinary.

Annular eclipses lend themselves very well to hydrogen-alpha telescopes primarily because they can show prominences. In fact, during the partial stages preceding a total eclipse, having such a telescope removes the element of surprise that used to be derived from unexpected prominences showing up suddenly when totality begins. Now, observers can know precisely how many prominences there will be and where they appear.

### Science in annular eclipses

Since an annular eclipse is a partial eclipse, the amount of scientific study that can be accomplished during one is quite limited. There are always meteorological phenomena to be studied, however. During the annular phase the temperature will drop; it is interesting to keep notes of this, as well as a visual record of any clouds that might form, or dissipate, as the eclipse progresses. As temperatures decline, the amount of humidity in the lower atmosphere will affect the formation (or dissipation) of clouds.

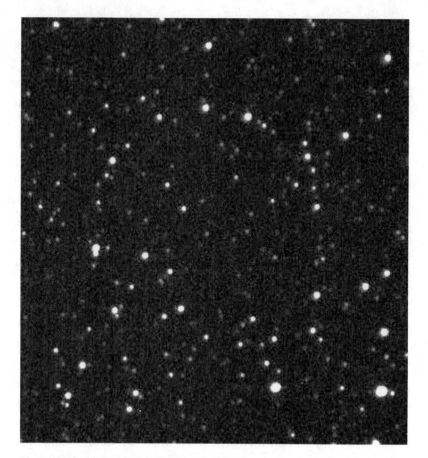

**Figure 6.5** Although it is difficult to find features in the sky with the same annular diameter as the Sun, this almost perfect circle of faint stars does the job nicely. Informally called Wendee's Ring, it is an asterism of unrelated stars discovered on January 2, 2000, by Wendee and David Levy. Photograph by David Levy.

Astronomers have conducted more advanced studies of Earth's ionosphere and thermosphere using "incoherent scatter" radar observations. In this observation, a radar beam bouncing off electrons in the upper atmosphere – the ionosphere – creates a measurable signal which is used to record the density of electrons, temperature of the ions, the composition of the ions, and their velocity.[2] The Millstone Hill Observatory uses the radar to measure the Earth's ionosphere.

As recently as the annular eclipse of September 23, 1987 through Japan, I. Sato and M. Soma continued their long-term study to see if the diameter of the Sun changes ever so slightly over the years. "One of the most precise methods to determine the solar radius," they claim, "is the observation of grazing solar eclipses. By the observation of the grazing solar eclipse at the southern limit line which passed on Kerme Islands near Okinawa on 23 September 1987, the author determined the solar radius as 959.64+0.37 ±0.10, which is consistent with predicted values and values obtained at earlier eclipses.[3]

David Dunham of the International Occultation Timing Association has also worked in this area of research. He and others deliberately avoid traveling to the eclipse center line, thereby missing most of the total phase. Instead they witness a "flash" of totality during which the corona might appear for an instant and then vanish. They did record a small increase in the diameter of the Sun, about 180 miles, between the eclipses of 1979 and 1983.[4] Thus, it appears that the Sun's diameter is pulsating slightly, and this can be measured during total eclipses, and during annular eclipses that are nearly total.

## ENDNOTES

[1] D. Fischer, First impressions, http://www.astro.uni-bonn.de/~dfischer/aus99/story.html. 20 July 2009.

[2] Millstone Hill Observtory Incoherent Scatter Radar Tutorial, http://www.haystack.edu/atm/mho/instruments/isr/isTutoril.html/18 June 2009.

[3] M. Soma and I. Sato, Observation of the southern limit of the annular eclipse in Okinawa on 1987 Sept. 23, *Twenty-first Symposium on Celestial Mechanics*, 1988, 101–107.

[4] F. Espenak, M. Littmann, and K. Willcox, *Totality: Eclipses of the Sun*, 2nd edn. New York: Oxford University Press, 1999.

## 7

# A total eclipse of the Sun:
# an introduction to the magic

*. . . dews of blood,*
Disasters in the sun, and the moist star,
Upon whose influence Neptune's empire stands
Was sick almost to doomsday with eclipse . . ."
(Shakespeare, Hamlet, 1.1.117–120)

What is the difference between a total eclipse of the Sun and anything else Nature has to offer? All the difference in the world. A total solar eclipse, it has been said, is Nature's grandest offering. Nothing – not even for me the discovery of a comet – can approach the ecstasy that accompanies the sight of the Sun vanishing and being replaced by a jeweled crown. It's not just the appearance of the Sun transformed that does it; it appears that the rapidly dimming crescent Sun, increasingly covered by the Moon, provides a real-time dynamic event that is truly breathtaking.

## The eclipse of 1918 . . . and of 2008

Of all the eclipses I have seen, for now I'd like to focus on just one – the eclipse of August 1, 2008. It was a repeat of the total eclipse that occurred 90 years earlier, on June 1, 1918. In a sense totality never ended that day long ago. Of course, totality could never last more than seven minutes and a few seconds for any eclipse, let alone a century, but in the minds of those who watch, the experience *never* ends. There is really no other way to write it; a total eclipse of the Sun is likely Nature's most incredible spectacle.

I first became aware of a total eclipse in 2008 while visiting Leo Enright, one of my oldest and closest friends. He outlined for me the moment when the Moon's shadow first struck the Earth in northern Canada not far from a Canadian Forces base at a high latitude in northern Canada. The shadow would then track north to a point not far from the Pole, then gently curve southward again through Russia, and end over Mongolia.

The history of this particular eclipse didn't strike until Wendee and I were at the Ob reservoir south of the Siberian city of Novosibirsk, and the eclipse had actually started. As the partial phase deepened I did a quick mental calculation and realized that I was enjoying the same eclipse that passed over the United States 90 years earlier on June 8, 1918, drawing a narrow track of totality through the heart of the United States, and producing a deep partial over a farm near Delphos, Ohio. I call this event the Leslie Peltier eclipse, recorded for posterity in his autobiography *Starlight Nights: The Adventures of a Stargazer*:[1]

"At mid-eclipse I turned away and looked about. Everything I saw, the nearby fields, the distant vistas, all seemed wrapped in some unearthly early twilight. The sky seemed darker – shadows faint and indistinct. A cool wind, almost chilly, had sprung up from the west. The grass beneath the nearby maple now was appliquéd with scores of crescent suns, projected there from each small aperture between the leaves above.

"Back again at the telescope I could see that now the darkest phase had passed. Seated atop my low stepladder I watched, fascinated, as the moon, now in full retreat, slowly relinquished all the solar spoils which it had won. From behind the low serrations of the profiled mountains of the moon, one by one the sunspots now emerged from occultation. . . .

"Along the narrow track of totality astronomers from all over the world packed up their precious plates and prepared to leave for home. Weeks before they had assembled here and had carefully taken their places in line in order to see a spectacle that would last just two brief minutes. For the most part they left well pleased with the performance though, as always, some had been unfortunate in their choice of seats along the lengthy aisle. And as they started homeward not one in all that far-flung audience could know that this was just an intermission and that the show they had come so far to see would be a double feature.

"When darkness came that evening I clamped my spyglass to the grind-stone mount which still was standing at the station underneath the walnut tree. I hoisted it up on my shoulder and carried it out to the middle of the front yard and stood it where I would have a clear view of the variable stars in the southwestern sky. That was the night I forgot all about telescopes

**Figure 7.1** The diamond ring viewed immediately after totality on August 11, 1999. Photograph by Roy Bishop, Acadia University.

and variables for as I turned and looked up at the sky, right there in front of me – squarely in the center of the Milky Way – was a bright and blazing star!"

That was it – on the evening of June 8, 1918, a few hours after a solar eclipse, light from a distant exploding sun arrived at Earth. When the nova reached its maximum brightness the following night, Leslie recorded its magnitude as –1.5, about as bright as Sirius, the brightest star in the sky.

I include these words here because I consider them the most moving thing ever written about an eclipse. Leslie Peltier's gem placed the eclipse of 1918, through the power of his wordcraft, among the great eclipses in history. I cannot imagine better words to inspire people to seek out and observe these great natural events.

**Figure 7.2** Totality in Antarctica, November 2003. Photograph by David Levy; however since his hands were freezing and the camera was not stabilized, the original showed a brief, perfectly exposed eclipsed Sun overlain by camera shake. The picture was corrected later by Keith Schreiber.

The eclipse repeated itself in 1936, and by a remarkable coincidence Peltier was observing the night before and discovered another nova, this one fainter, in the small summer constellation of Lacerta. Then came 1954, when a group of travelers from Montreal were clouded out. On July 10, 1972, I caught the maximum partial phase of the eclipse in Montreal on an otherwise humid and rainy day. When the eclipse returned in late July 1990, I observed a few minutes of the opening partial phase as the Sun was setting in northern California. And when it returned once again, on August 1, 2008, I captured a magnificent view of totality in Siberia.

It is difficult to describe in words everything that happens during a total eclipse. The entire duration of an eclipse usually exceeds two hours. but if the eclipse is total, the maximum phase never lasts more than seven minutes and can be as short as a few seconds. The total phase of the 2008 eclipse was a bit over two unforgettable, precious minutes.

In the early 1980s, Norman Sperling, then an editor at *Sky & Telescope* magazine, coined a now-famous catchphrase that all total solar eclipses last seven seconds. There is some validity in what he said. The moments of total eclipse are so exciting, so unnerving, that even during a long total phase the time races by.

The following two chapters will explore what observers can do during the total phase of an eclipse, as well as during the moments leading up to the onset of totality and the time immediately after totality ends. But in this chapter, let's capture the magic.

Most of us first read about total eclipses in books. Although there are plenty of photographs of total eclipses, I have yet to see a single one that comes close to capturing the visual experience. Even images that claim to capture what the eclipse looked like to the eye lose so much that they cannot really lay claim to having achieved what they set out to accomplish. The simple fact remains: no number or frequency of eclipse photographs can make up for actually seeing a total eclipse of the Sun.

## Travel is usually necessary

Total solar eclipses are often called "once in a lifetime events." They are, if you do not travel far to see one. For example, Montreal, Canada (one of the luckier cities) will enjoy a total eclipse on April 8, 2024. The previous total visible in that city (or at least the western portion of it) was on August 31, 1932. Our Jarnac Observatory near Vail, Arizona, is less fortunate. There has not been a total solar eclipse at Jarnac since August 26, 797 (though there was a brief 10-second total on May 16, 1379, and annulars in 1740 and 1821), and the next one will be about 1700 years later, on October 15, 2498 (not counting two annulars in 2269 and 2422). Throughout my lifetime, I needed to do some travel for every total eclipse I have seen. The eclipse of July 20, 1963, was not far from my Montreal home, although I was living in Denver at the time. The eclipse of March 7, 1970 took place just south of Acadia University, where I attended as a student at the time. The total eclipse of June 21, 2001 took place in the southern Africa country of Zambia. The farthest I ever have traveled (or ever expect to) was to the Novolazarevskaya science station in Antarctica, for the total eclipse of November 23, 2003.

Travel involves planning. If totality will occur over a major city, it is usually a simple matter to travel to that place, as we did for the annular eclipse over Madrid on October 3, 2005 or the total eclipse over Winnipeg on February 26,

**Figure 7.3** Totality in Antarctica, detail showing the eclipsed Sun, the corona, and an exultant Japanese observer enjoying the eclipse. This was a difficult shot that captured both the eclipse and the thrill of viewing it. Photograph by Fred Bruenjes.

1979. More often one has to fly to a city, then travel by bus or car to an outlying area in the path of totality.

Traveling a great distance to view an eclipse also involves planning for the weather. Take Jarnac Observatory, for example. The eclipse that took place on an August afternoon in 797 was in a place with a high possibility of cloud cover due to the summer storm season that is prevalent at that time of year. Never mind that there won't be another total eclipse for almost 2,000 years; whatever the weather happened to be that day would have taken precedence. And when the next total eclipse of the Sun takes place over our observatory in the fall of 2498 (during a particularly clear time of year), the sky may or may not be clear during that time (unless we have a weather modification net in place to take care of any clouds by then).

Certainly by 2498, there will be very little science to be done from the ground, even during the total phase of an eclipse. Professional astronomers will not request grants to travel far to view such an eclipse. But the wonder of the event, the emotions, will remain undiminished, no less than it was during the eclipse of 797. People will gather, prepare for the event, and even photograph it using whatever advanced equipment is available at the time. The science may be gone; the poetry remains.

Before the days of coronagraphs and spacecraft, totality offered the only opportunity to learn about the Sun's atmosphere and its relation to Earth and to the Universe. Although we can now view prominences at any time, seeing them during a total eclipse adds significantly to their magic.

## Science in total solar eclipses

The excellent adventure of Eddington and the eclipse of 1919 has already been told in Chapter 2. But just how firm was the evidence? Considering the equipment available for this experiment so soon after the First World War, it is remarkable that the team was able to obtain any results at all.

The eclipse of 1919 was not the first eclipse to yield scientific data about the Sun. There is some evidence that the experiment was conducted as early as the "Peltier" eclipse of June 8, 1918. According to authors Littman, Willcox, and Espenak, two American astronomers, Heber Curtis and William Campbell, led an expedition to Goldendale, Washington (not far from the site of totality for the 1979 eclipse decades later) to measure the deflection of stars near the Sun. Unfortunately, the Sun was not that close to a bright star cluster at the time, and clouds during part of totality obscured some of their images. Although some of their plates had stars in them, there was a focus problem that really prevented accurate measurements of the stars' positions. Thus, the 1918 results were inconclusive.

The famous experiment was repeated at many eclipses, until 1973 when a series of 150 measurements during a relatively long period of totality – more than six minutes – confirmed Einstein's prediction more accurately than ever before. Although other methods, using radio telescopes and quasars, can now measure the light deflection anywhere and at any time, the 1919 solar eclipse was the first real test of Einstein's theory of general relativity, and its success properly catapulted Einstein to the top rung of the ladder of modern physics.

Total eclipses in the past have yielded fascinating information. The eclipse of 1605, which was total in parts of Europe, somehow suggested to Johannes Kepler, who even missed totality, that the corona was made of light reflected from circumsolar particles. (If only he knew how close to being right he was).

What eclipses have taught us date a lot further back than that. In 1307 BCE "three flames ate up the sun" during an eclipse. In 450 BCE Anaxagoras proposed that the Moon is lit by the Sun. In 648 BCE Archilocus, swayed with amazement at a solar eclipse, called it "inexplicable," a description that bears weight even today; understanding the mechanics takes nothing away from the magic.[2] However, Thales, it is said without firm evidence, might have predicted the eclipse of May 28, 585 BCE.[3] That event was total in parts of Italy and Greece. If Thales did so, he probably holds the record for the earliest person to predict an eclipse, which would have necessitated a good understanding of the fact that the Moon passing in front of the Sun causes one.

### Anaximander and the Sun

The famous Greek philosopher Anaximander has an original theory of eclipses, since they dovetail with his father's proposal that the Sun, Moon, and the other stars are fires contained and restricted in spherical regions surrounded by cool air which maintains them. We are able to see these fires through vents or shafts like those found on a musical instrument. Should any of these shafts be blocked, an eclipse results.[4]

Stories and theories like this one show the richness of thinking about eclipses over time, and that serious people have considered these events over time. It does not matter whether these theories were correct, or even approached correctness; ideas like these show the variety of thought that spanned the world about eclipses over the ages.

### Thoughts

No matter what else you try to do during the total phase of an eclipse, spend a little time feeling its magic. The eclipse is a masterwork of celestial complexity and precision. It's certainly not just the Sun; even if the sky is completely cloudy your view of the sudden sky changes is remarkable; in fact a clouded out sky during a total eclipse. As evidenced by our view of the 1970 eclipse from Nova Scotia, is usually darker than a clear sky view.

Finally, it is important to remember that every eclipse is different. The sky has a slightly different appearance for each eclipse; not only that, but also each observer will have a slightly different description to offer about the sky for a particular single eclipse.

## The emotions of totality

After reading Leslie Peltier, it seems odd that we would follow it with a section on the emotions generated by a total eclipse. But as you prepare for one, or as you relive one, it helps to get your emotions in order. As totality approaches, feelings get churned up. A large group has a tendency to give loud voice to these feelings.

Why is the experience of totality so poignant? I believe the answer to this difficult question lies rooted in humanity's past. In ancient times, some viewers really feared that the Sun was disappearing forever, and this sense of panic is not surprising. During the February 26, 1979 eclipse, I found myself wishing that the rush to darkness would slow and halt, even for a moment. I had these feelings even though I know exactly what causes a solar eclipse. Even so, it was interesting to share them.

During that eclipse I was part of a very small group of four. We were virtually silent during the total phase, save for single word remarks like "Prominences!" On the other hand, large groups of people tend to get noisy, even rowdy, as the onset of totality approaches. During the eclipse of 2008, we were part of a crowd of thousands of people assembled on the beach at the northern end of the Ob reservoir. At four minutes before totality you could hear the excitement level climb. Venus popped out at about 3 minutes 15 seconds before totality, "at 10 o'clock slaunchwise from the Sun." Shadow bands along the beach appeared at two minutes prior to totality. With the beginning of totality the screams were almost deafening, but they rapidly subsided as observing programs set in. Mercury bowed onto the celestial stage as well, in the same direction as Venus but much closer to the Sun.

Such unbelievable cheering and noise does not usually accompany a phenomenon of Nature. That it does during a total solar eclipse testifies to the awe and inspiration that such events bring.

ENDNOTES

[1] L. C. Peltier, *Starlight Nights: The Adventures of a Stargazer.* Cambridge: Sky Publishing, 1999.

[2] Dirk L. Couprie, Robert Hahn, and Gerard Naddaf, *Anaximander in Context: New Studies on the Origins of Greek Philosophy.* Albany: State University of New York, 2003, 31.

[3] *Ibid.*

[4] http://www.thebigview.com/greeks/anaximander.html. Accessed 8 July 2009.

# 8

## The onset of totality

In the last few minutes of a partial eclipse that will become a total eclipse, a lot of things begin to happen. The sky's gradual darkening becomes more sudden, and in the final moments the sky darkens as quickly as if someone were working a huge celestial dimmer switch. From the west, or northwest, or southwest – wherever the shadow is coming from – the sky is blackening even more rapidly. Clouds might still be visible even though there is very little sunlight left to fall on them. The Sun itself appears as a crescent so thin that it appears only as a line.

### Venus appears

At almost every eclipse, between ten and seven minutes before totality begins, someone will yell a single word: "Venus!" Everyone then strains to catch a glimpse of Earth's hellish sister planet. Venus is a symbol. The sight of our neighbor world means that the sky has darkened sufficiently for the brightest planet to appear along with the crescent Sun. Along with the appearance of Venus, shadows are becoming very sharp and distinct. Just prior to totality during the eclipse of April 2005, Venus appeared due north of the Sun. The planet was near its conjunction; less than a year earlier it actually transited the Sun for the first time in more than a century.

### Shadow bands

One of the most intriguing aspects of a total solar eclipse is the shadow band effect, which happens when thin wavy lines of darkness appear

**Figure 8.1** In the vicinity of Lundar, Manitoba, the sky darkens rapidly as the total phase of the 1979 eclipse rushes in from the southwest. The right half of the picture is darker than the left, indicating the shadow's direction of travel from the southwest. Photograph by David H. Levy.

just before or after totality. They usually begin very faintly, then intensify as totality approaches. During totality they vanish, but often return in force, and pretty stongly, immediately after totality. Then they fade gradually.

By far the best display of shadow bands I have ever seen was during the November 23, 2003 eclipse in Antarctica. From my own observing log, session 13680S: "At about 7 minutes before totality I discovered faint shadow bands in the sunlit area of snow in front of us. They were quite regular waves of dark lines moving at about a meter per second away from the eclipse. As totality drew near, the bands darkened and grew more obvious." I do recall seeing the bands again after totality, and the total time for the view was about 12 minutes.[1]

Wendee added to the description, reprinted from her Journal: "It looked like a ceiling fan was moving because the white snow had black smoky lines running across. As totality neared, the shadow bands were more vivid and were still that way after totality ended." Her eloquently chosen word to describe these particular shadow bands was "strobing."[2]

The best time to begin looking for shadow bands is about six minutes before totality commences. To me, the best analog is the shimmering of light

**Figure 8.2** Totality on February 26, 1979. This Ektachrome 200 picture concentrates on the darkened Sun and bright corona, leaving out the prominences. Photograph by David Levy.

around a swimming pool. They offer a unique, unearthly effect. Add this check to your already hectic schedule of observations of the shrinking crescent and the onrushing darkness of the Moon's shadow. Check again around four to three minutes prior to the start of totality. This time, your chances of seeing the bands are considerably greater. If the site you are using is not flat and brightly colored (snow or sand), placing a white sheet on it should help. I think that the reason we saw shadow bands so well in Antarctica was that they presented themselves upon endless miles of pure white ice and snow.

During the middle part of the last century, even amateur groups were organized with the purpose of making some scientific contribution during an eclipse. Thus, during the eclipse of March 7, 1970, a group from Montreal concentrated on observing shadow bands, making detailed temperature records, and other items.

As totality approached, Donald Frappier and Allan Moore used a large sheet variously colored white, grey, and black. "Fifty to sixty seconds before [the onset of totality], a hint of shadow bands noticed by myself and partner were very, very faint. Positive identification 36 seconds before and lasted into totality for six seconds. Shape wavy, four inches apart, white section between shadows whiter than the white part of the sheet, showed as white on grey section, and very faint white on black section. Thickness of shadows approximately $\frac{3}{4}$-inch. Frequency – two or three bands passed a given point every second. Direction of travel – forty-five degrees from north to east. On a scale of zero to five, intensity approximately two. Rate of travel constant, smooth."

Apparently this group was looking for shadow bands during totality, thereby missing some precious seconds of corona to do it. They indeed recorded bands stretching about five or six seconds into total eclipse, and then did not see any until about five seconds after the end of totality, when they "seemed to start very abruptly. No doubt as to beginning. All of a sudden they were there, very prominent. Lasted about twenty seconds. Shape seemed different from that before totality, more straight than wavy. Also, there seemed to be a wide shadow and a narrow shadow separated by a white section. About two inches apart. Intensity 4 on a scale 0–5. They were flickering like old movies used to flicker. Direction of travel thirty degrees north to east. Very intense, much more so than before totality. Plainer on sand at feet than on sheet." This remarkable report appeared as part of the Montreal Centre's newsletter *Skyward*'s coverage of how its own members traveled to see the eclipse.[3]

All these effects our own group missed that March day in 1970, since it was clouded out at totality.

### Baily's beads

The group you're with, particularly if it is a large one, is probably screaming with excitement and wonder, and you're spending these opening seconds trying to get in as much as possible. But as the seconds tick away, the screaming dies down and it's time to get on with your program.

As the solar crescent diminishes to a line of light, that line of sunlight shrinks further until it begins to break up. Overhead, sunlight is shining only through valleys at the limb, or edge, of the Moon. This effect, which lasts no more than a few seconds, is credited to Francis Baily at the eclipse of May 15, 1836. Although other observers may have seen the beads earlier, it was Baily's intricate description of the Sun's shining through lunar valleys at the limb of the Moon that has forever identified his name with this phenomenon. Baily was lucky; the eclipse of 1836 was a very brief annular eclipse over Inch Bonney, Roxburghshire, Scotland. Baily apparently reported on his beads at the Royal Astronomical Society meeting of December 9, 1836.[4] This effect allowed Baily to observe the beads in enough detail to describe that "a row of lucid points, irregular in size, and distance from each other, *suddenly* formed round that part of the circumference of the moon that was about to enter on the sun's disc."[5] Baily had read of earlier observers, like Edmond Halley, who had noticed the beads. This nineteenth-century astronomer used a $2\frac{1}{8}$-inch-aperture refractor, and a big 20-inch prismatic telescope designed to measure the angular distance between the Sun's limb and that of the Moon, two thermometers, a burning-glass, and four pocket watches.

**Figure 8.3** A view of the total eclipse of the Sun, February 26, 1979, showing the inner portion of the corona and some gorgeous prominences. Photographed using a 3.5-inch Celestron telescope (now named Pumpkin) by David H. Levy.

Baily continued his observations of the beads, and was surprised to discover that "these luminous points, as well as the dark intervening spaces, increased in magnitude; some of the continuous ones appearing to run into each other like drops of water."[6] Since the 1836 eclipse was annular, the beads evolved into a continuous arc of light as the annulus formed. Baily's beads, as they are now known, are staples of total eclipses as well. However, their performance differs during a total eclipse. Instead of running into each other as in an annular eclipse, the beads get smaller and quickly disappear until only one is left. As beautiful as the Baily's beads effect might be, it is fleeting, lasting at most a few seconds unless the observer is at the very edge of the path of totality.

## The emergence of the corona

While Baily's beads are visible, in fact sometimes a few seconds before they appear, the corona begins to appear, faintly at first, and more obviously on the side away from where the beads are forming. The size and shape of the corona varies from eclipse to eclipse. There appears to be a relationship between the shape of the corona and the status of the sunspot cycle: the closer to sunspot maximum, the more oval the corona. Near the sunspot minimum the corona appears almost circular; it becomes more oval as the cycle approaches maximum.

## The diamond ring

As the corona begins to appear, Baily's beads rapidly disappear until just one is left. For a second, the Sun is completely covered by the Moon, save for one spot near the center of where the thin crescent was a moment earlier. A brief look at the Sun will show the emerging corona accompanied by the single bright spot. The entire effect is called the diamond ring. It is one of the most stunning, unbelievable sights to see during an eclipse.

At the 1999 eclipse, Wendee was taking a series of photographs through her simple point-and-shoot camera. One of the staff members of our cruise was conducting a countdown of sorts, but he was off by a minute. Around the time he said there was one minute left, suddenly, through her camera she saw the corona begin to form and that the crescent had reduced to a single bright spot of photospheric light. "Diamond ring!" she called loudly. She was

The ECLIPSE-O-SCOPE to VIEW the SUN

WEDNESDAY AFTERNOON, AUGUST 31, 1932

(For exact time see local newspapers)

Copyright 1932
Art Photo Co.

Non-inflammable
Filter

Made by
Art Photo Co.
Springfield,
Mass.

A Solar Eclipse is the most beautiful and inspiring sight of all the natural phenomena. When the sunlight is noticeably decreased, the time of totality is near. The interesting changes to look for are peculiar color effects, light streamers from the sun, and wavy shadow bands closely followed by the swift travelling shadow of totality, and round beads on the edge

of the disk, called Bailey's Beads. Almost immediately occurs the crowning glory of the spectacle, the bursting forth of the beautiful pearly light of the Corona, a sight unequalled for its wonderful grandeur.

The Harvey & Lewis Co.

OPTICIANS
KODAKS        MOVIES
Hartford, Conn.    New Haven, Conn.
Bridgeport, Conn. New Britain, Conn.
Springfield, Mass.  Worcester, Mass.

**Figure 8.4** In 1932, the optical Harvey & Lewis Company manufactured this "Eclipse-O-Scope" to view the Sun safely during the eclipse that afternoon. Photograph of this antique by David H. Levy.

so excited I feared she might jump right off the ship. This was Wendee's diamond ring – she had discovered it, seen it before anyone else; with the Sun so close to the horizon she might have been the first to see it all along the track. The diamond ring is the gate into totality; eye protection comes off, and for the next 50 seconds it would be possible to gaze at the entire spectacle without any protection at all. "Glasses off, everyone," was what I stated in response to Wendee's alert.

Is it dangerous to look with unprotected eyes at the diamond ring? Yes. The diamond is a spot of brilliant sunshine that has all the power of full sunlight to damage the retina. In my experience, however, the incoming diamond ring lasts only about a second before it fades away to full totality. It is far *more dangerous* when it appears again at the end of totality. My advice: do not remove your safety glasses until Baily's beads are dissolving into the diamond ring. This procedure will give you a glimpse of the ring just before totality starts. When totality ends with a second diamond ring, take one glimpse of it and then get your eye protection back on immediately! It gets only worse from there. But once your glasses are on again, look again at the Sun to capture Baily's beads.

2 pm October 6, 2005, Barcelona.  At 2:04 pm on October 12, 1605, a total eclipse of the Sun appeared in Barcelona.

**Figure 8.5** Barcelona on October 6, 2005, just 400 years after a total eclipse appeared there. Photograph by David Levy.

## Prominences

Until recently, the thin flame-like strands, called prominences, that appear to leap off the limb of the Sun were a complete surprise reserved for total eclipses. But the current availability of inexpensive telescopes using hydrogen-alpha filters means that in almost any group of eclipse watchers, at least one will probably have such a telescope and will be able to advise the others where to expect prominences. Again, near sunspot minimum, fewer prominences should be present than when the cycle is near its maximum. At that time an eclipse might feature a dozen large prominences or more, and these features only add to the magnificence.

Prominences behave like another type of solar storm called sunspots; they follow the solar activity cycle, which changes sinusoidally, ranging from minimum to maximum and back again every 11 years. A typical prominence appears as a thin line (straight or curved) of hydrogen rocketing out and away from the Sun's photosphere. A prominence is not easily visible until it is at least the size of Earth, about 8,000 miles long. A particularly strong prominence can thicken and stretch hundreds of thousands of miles away from the photosphere, and even though such a prominence is remarkable through a hydrogen-alpha filtered telescope, to see one visually during a total eclipse is gorgeous quite beyond description.

### Evaluating the sky

Depending on the length of totality, and the observer's position within the Moon's shadow, the sky will display many changes during totality. I have already described the rapid motion of the Moon's shadow across the sky (like the second hand of a giant clock) during the 1999 eclipse. But totality during that eclipse, with the Sun so low in the sky, was less than a minute as the Moon's shadow tore along at some 12,000 miles per hour. Most total solar eclipses last two minutes or longer and allow changes in the sky to occur a little more slowly.

First, the brightness of the sky changes rapidly during totality. As totality begins, the west, northwest, or southwest is the darkest part, and during the course of totality that darkest spot shifts as the shadow passes over. It is usually easy to see the edge of the shadow to the south or north. At mid-totality it is possible to look all around and see the shadow's limits, the line that divides totality from very deep partiality, twilight from daylight. Some observers, particularly those mentioned in the chapter on annular eclipses, deliberately position themselves at the *edge* of totality specifically to extend the precious moments allotted to viewing and measuring features like Baily's beads as they appear to swing around the Sun.

### Studying the sky

During a total eclipse the sky changes rapidly and drastically.

What planets are there? At every total eclipse you should be able to spot Mercury and Venus. Venus actually appears in the last few minutes before totality begins, and usually lasts for several minutes after totality ends. Mercury is more of a totality-only companion, but seeing it at all is a special treat since most professional astronomers, actually, have never seen Mercury.

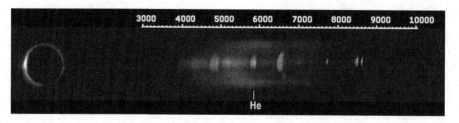

**Figure 8.6** During the 2006 eclipse in Turkey, Robin Leadbeater captured this view of the flash spectrum of the chromosphere. You can see it for a few seconds while looking through an objective prism, just after totality ends. Photograph by Robin Leadbeater, used with his kind permission.

*Play of light and shadow*: Look around the sky. Which portion is darkest? The shorter the length of totality, the faster the Moon's shadow will appear to swing across the sky. During the April 2005 hybrid eclipse, our group was in the middle of the brief section where the eclipse was total, but only for 29 seconds. As our ship cut its way through a rough sea, the shadow virtually tore its way across the sky from the northwest to the northeast. On July 11, 1991, the shadow of the Moon strolled majestically from the northwest to the southeast during almost seven minutes of totality.

No matter how brief or extended totality is, note the condition of the sky. At the start of a long totality (four minutes or more), the sky should still be fairly bright, but as the time moves on, it darkens further and more quickly. The shadow is thickest at its center, which happens to be where the Sun is. So you really never get a sense of just how dark the Moon's shadow can be.

Are any bright stars visible? During the brief 50-second totality in Antarctica during the eclipse of November 2003, I could spot some of the sky's brighter stars like Sirius, Canopus, and possibly Achernar. I also checked the sky for aurora australis, which could have been visible had conditions been right. But except for a few stars and planets, the sky was pretty quiet.

### Any comets?

In the years before spacecraft like the Solar and Heliospheric Observatory (SOHO), HELIOS, STEREO and the Solar Dynamics Observatory monitored the region around the Sun, it was possible to discover a new comet during a solar eclipse. In 1882 and in 1948, new "eclipse comets" were discovered during the few moments of totality. But these days it would be highly improbable to find a comet during an eclipse that has not been seen before.

Things were pretty different in November 1948, six months after I was born. As the lunar shadow crossed over Africa, the Indian Ocean, and south of Australia on November 1, 1948, a brilliant comet, perhaps magnitude –2, shone just two degrees south of the eclipsed Sun. The comet reappeared on the morning of November 4 in the morning sky but was visible only from the very lowest northern hemisphere latitudes, and from the southern hemisphere sky.

That great eclipse comet was not the only comet discovered during a solar eclipse. On May 17, 1882, a group of observers enjoying a solar eclipse in northern Egypt spotted what appeared to be the tail of a comet stretching away from the Sun. They even had a photograph on which the comet appeared clearly. The group decided to name the comet "Tewfik" which is not in accordance with any established protocol for naming comets, but which is informally used for the comet that was spotted but once, during the eclipse.

Although it is highly unlikely, it is not impossible that a bright comet will appear during a total eclipse. Such a comet would probably be very close to the Sun, within a degree of it. If you spot such a comet, be sure to report it as soon as you can reach a phone or a computer. In your report, specify as accurately as you can where the comet was relative to the Sun, its brightness, length of its tail, and the direction in which the tail is pointing. Also, if the tail curves as it spreads out, be sure to mention the direction to which the tail is curving.

Even if you do find a comet during totality, your attention will be so much drawn to the corona and other aspects of the total phase that it may be difficult to recall much about the comet at all. That's why it is important that someone have a photograph. It is likely that the comet has appeared on a photograph, which means that someone can measure it far more accurately than you could visually.

This may be the only eclipse guide that suggests you look for undiscovered comets during totality, and with good reason – its author is passionate about comets. I have spent at least a little time during each totality scanning the sky near the Sun for comets. I don't spend much time on this because spacecraft have all but taken over the discoveries of comets near the Sun, and because they are necessarily in space, they can find these comets anytime – whether the Sun is in eclipse or not.

**Figure 8.7** A remarkable photograph of the 1999 total eclipse of the Sun taken with a simple "point-and-shoot" camera. Note the detail in this picture, which captured the Moon's shadow and its edges on either side of the Moon. Photograph by Wendee Levy.

## Entering the central portion of totality

Much of what we have discussed to this point takes place in the opening seconds, or at least the first minute, of totality. The experience is now settling in. The Sun is totally, completely covered by the Moon. It is time to move on to what you plan to do during this critical phase.

ENDNOTES

1 Observing Log, Vol 20.
2 Wendee Levy, Journal, November 23, 2003.
3 More reports on the solar eclipse of March 7th, *Skyward*, May 1970, 6.

4 Francis Baily, Communication to *Monthly Notices of the Royal Astronomical Society*, 4 (2), 15–19.
5 Baily, 16.
6 Baily, 17.

## 9

# Observing a total eclipse of the Sun

Methinks it should be now a huge eclipse
Of sun and moon, and that th'affrighted globe
Should yawn at alteration.

(Shakespeare, *Othello*, 5.2.108–110)

An observer with much experience with eclipses, Steve Edberg, gazed silently at the Sun when its corona appeared near midday on July 11, 1991. Just stared. Even though he had seen the corona several times before, he was overwhelmed with its size and beauty. This was to be a long total eclipse – almost seven minutes of totality – so he had the chance to preserve this moment. "Well now," he muttered after the moment. "Perhaps I should take a picture or something?"

No matter how well you prepare for it, unless you are a person of extreme discipline, the onset of totality will drag you away from your planned program. You just want to gaze at it. That is actually the right approach, and no matter how much or how little time totality will last, you do want to have some time for just taking it all in. If totality is long, as it was in 1991 and in 2009, then you have some time for this. If totality is short, then make the time. The shortest totality I've ever experienced was the hybrid eclipse, annular in places and total in other places, in April 2005. Even then I gave myself about five of the 29 seconds of totality to enjoy the view before I resumed my plan.

### Looking around

The landscape or seascape provides one of the most interesting views during totality. Look around you. Are there any animals around? They may

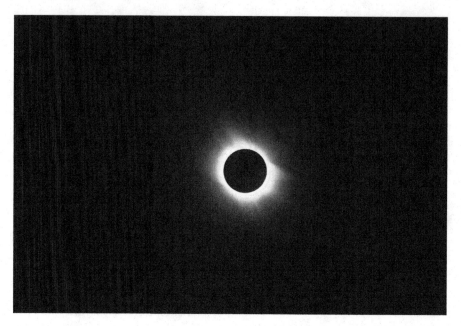

**Figure 9.1** A view of the total phase of the August 1, 2008 eclipse. Photograph by David H. Levy, and enhanced by Dean Koenig using Adobe Photoshop.

be quietly preparing for nightfall. Birds may flock to their nests. If you gaze towards the horizon, you should see it as quite a bit brighter all around you, for the horizons are outside the shadow. Light from the boundaries of the shadow helps to brighten the sky within the zone of total eclipse.

What is the temperature? If you are so inclined, record it at the beginning and again just after the end of totality. It should have continued to drop sharply during the total phase, and should start picking up shortly after totality ends.

What about the wind? Although I have read nothing, nor have I uncovered any evidence that suggests that an eclipse can change the weather, we do know that a total eclipse causes the temperature to drop quickly. This could result in the wind velocity increasing or decreasing. But during the Antarctic eclipse of 2003, the wind picked up sharply a few minutes before totality, compounding the already bitterly cold temperature of –23 degrees Celsius. After totality the wind did not subside; in fact several hours later it was still strong.

### Details of the corona

The major questions regarding the corona concern its shape and color. As already explained, eclipses around sunspot maximum tend to produce a circular corona, and it becomes increasingly oval as the solar cycle approaches its minimum. The corona is usually white, sometimes called pearly white (although those descriptions are virtually synonymous), or even silvery. Virtually all photographs show the corona as white.

Look *carefully* for structure within the corona. The Sun has a complex magnetic field which the corona closely outlines. Large streamers may emanate from the inner part, stretching some distance (up to two degrees or more) out from the Sun. While photography can help provide details of the corona, drawing is also a useful thing to try. In his sketch of the corona on June 21, 2001, Stephen James O'Meara noted many intricate structural details, and his keen eye and accurate pen helped reveal these particulars.

## Prominences

As the Moon moves across the Sun, different prominences might reveal themselves. The "Seahorse prominence" from the 1991 eclipse, for example, did not appear until near the end of the long totality, and then it gradually grew to reveal its full dimensions, staying in full view until the diamond ring at third contact washed it away. Like all prominences, this one displayed a deep red color, contrasting nicely with the white of the corona. There was also a lovely set of remarkable prominences at the 1979 eclipse. However, even though the solar cycle should have permitted them, I did not see any prominences during the eclipses of 2003 or 2006.

## The chromosphere

The Sun has a thin atmospheric layer, about 1,500 miles high, located just above its bright photosphere. This layer is called the chromosphere. Even though its color and dimensions, completely surrounding the Sun, should make its appearance obvious, it is actually difficult to see. I made a specific attempt to see the chromosphere during the December 2002 eclipse. Thanks to interference by a layer of thin, low cloud, the corona appeared only faintly. However, the chromosphere was pretty evident, on one side of the Sun at the start of totality and at the other at its end. The chromosphere is not a solid ring of red. It is made up of thousands of spicules, skinny line-like tongues of gas that climb off the photosphere.

## The flash spectrum

This observation is one of the most useful things that one can do during a total eclipse. Spectroscopy is useful because it allows us to learn things about a distant point of light. From a star's spectrum we can classify the star's type, and hence we can derive its distance, size, and even its age. Spectroscopy, especially for a group of amateur astronomers watching an eclipse, is somewhat esoteric, but therein lies its fun.

It is surprisingly easy, though tricky, to observe the flash spectrum. All you need to do is place a slide containing a diffraction grating over your observing eye. As the chromosophere appears seconds before the diamond ring, the flash spectrum should appear, and it should stay until the diamond ring dissolves into Baily's beads. This experiment is not without danger, however, for your eye is unprotected while you are gazing at the flash spectrum. To see or photograph it through a telescope, it is necessary to place a diffraction grating somewhere in or on the optical path of the telescope. This way, as the light from the eclipsed Sun enters the path, it will be divided into its component colors. At the incoming or outgoing diamond ring, it is possible to get a spectrum of the chromosphere that reveals the emission lines from both hydrogen (the hydrogen Balmer series of hydrogen-alpha, -beta, and -gamma) and helium. In 2006, I placed a small grating over my eye in the last few seconds of totality, and saw essentially what the British amateur astronomer Robin Leadbeater's breathtaking photograph also revealed. I saw the Sun at left, including the chromosphere, and the same emission lines that the photograph captured.

As easy as the flash spectrum is to observe, understanding it is more complicated. It is mostly an emission spectrum, showing the emission of various colors from the solar limb at the moment of the beginning of totality and at the ending. It is possible to see the flash spectrum because the chromosphere is so thin. Also because it is so thin it is visible only for a few seconds at the beginning, and another few seconds at the end, but on the opposite side of the Sun.

The spectrum is called a "flash" spectrum because it is visible so briefly. During it, however, a careful observer can see emission lines of helium that were discovered during the eclipse of 1868. It is really the spectrum of the chromosphere, which is deep red because of the strong emission of hydrogen-alpha light.

## Returning to the corona: observing its spectrum

Although the spectrum of the Sun's corona lacks the specifics of the "flash spectrum" of the chromosphere, it is worth exploring. Its lines are not sharp, but they can be measured. Perhaps the most interesting is a green emission that, after many years, was finally determined to be caused by ionized atoms of iron each with 13 electrons removed. This discovery, by the Swedish astronomer Bengt Edlén, was a key factor in suggesting that the corona was much, much hotter than the photosphere.[1]

The Sun has three coronas. The first one is the K-corona (from the German word for continuous), visible to us during an eclipse because sunlight bounces off free electrons. There is also an E-corona (the E stands for emission), which is seen by emission lines from ions. Sunlight also bounces off particles of dust, enabling us to see the F-corona (the F stands for Fraunhofer). The F-corona extends far out into space where it evolves into the zodiacal light.

## Structure in the corona

The corona of the Sun is huge. During a solar eclipse it might stretch two or more solar diameters, but it is far larger than that. Carried along by the solar wind, it really extends well beyond the orbit of the Earth. The Sun's strong magnetic field dictates the shape of what we will see in the corona during an eclipse. The lines and shapes that turn the corona into a beautiful painting, hanging in the sky, are all related to the Sun's magnetic field.

*Coronal loops*: These major structures in the Sun's corona can be dramatic in totality. They are one of the major identifiers of the Sun's magnetic field. A loop structure often encloses a coronal hole, a dark area within the corona. Although a coronal hole may be seen anywhere around the Sun when totality is near solar maximum, such an event is more likely confined to the polar region during minimum.

*Streamers*: The corona is rarely perfectly symmetrical; more often one can see streamers coming out from the Sun. One popular type is called the "helmet streamers", bright loops which can develop right over active sunspot groups. Once formed, the solar wind stretches them to long oval shapes with sharp ends. An observer can expect them more often at eclipses that occur when the solar cycle is in a strong, active phase.

Helmet streamers are capable of evolving into coronal mass ejections or CMEs. If a large amount of plasma escapes from the top of a streamer, it can trigger a CME. These CMEs are major solar events. In 1859, the English astronomer Richard Carrington observed a major solar flare which led to a CME. Less than two days later, the CME reached Earth. It was accompanied by a major auroral display which was seen as far south as the Caribbean Sea. In those days, it was the world's telegraph system that was severely affected. When another storm developed in the spring of 1989, the resulting electromagnetic interference shut down satellites, damaged cell phone networks, and caused other damage. The 1989 geomagnetic storm produced auroras as far south as Texas, Arizona, and California, and changed the orbit of the Solar Maximum Mission satellite, launched by the shuttle and repaired in space a few years earlier. SMM's orbit was so altered by the March 1989 storm that the satellite tumbled and disintegrated in the Earth's atmosphere that December.

Although I do not recall a CME being visible during a total eclipse, the precursor to one was spotted during totality in Iran in August 1999. The underlying high-arch system (which resembled a dome-like structure with cavities inside) which appeared on the eclipse white-light images, was suggested to be a large-sale precursor of the CME. People casually watching the corona would not have seen this effect. The corona was unusual in that it contained several arch-like structures, one of which consisted of a number of arches. The observers used a radial filter "slightly shifted with respect to the Sun." There was a "high bright arch" inside which is a smaller arch also with its own interior cavity. Just under ten hours later, the underlying prominence erupted into a CME that, after a lapse of two more hours, was recorded on SOHO's LASCO 2 and 3 cameras.[2]

This report is a good example of how a detailed observation of a portion of the corona can yield important science; in this case it predicted a major solar ejection by just a few hours. I do not believe that anyone has seen a CME in progress during totality, although it is possible that one has occurred at least once in all the total eclipses that have hit the Earth over time. If one did occur, it might look like a quick "puff" of coronal material off the Sun in one direction in space. Space missions like SOHO show them erupting virtually instantaneously, so even if one were to occur it could be easily missed.

Whether it ends in a coronal hole, streamer, or other feature, the corona that we see during an eclipse is the place where the solar wind begins. That wind travels throughout the solar system, eventually stopping only when it comes into contact with the galactic wind, the combined energy from other stars in our galaxy. When we see the edge of the corona during a total eclipse, we can see the awesome division between the outreach of the corona and the real start of the solar wind that pushes far out into the solar system. This border is one of the most glorious sights to see during an eclipse.

## Diamond ring at third contact

In the last few seconds of totality, the sky from the west brightens dramatically, and so does the west side of the Sun. The chromosphere becomes briefly visible again, then finally the first bead of photosphere explodes in the darkness as the outgoing diamond ring. It is very dangerous to continue gazing at the Sun for more than a second or two after the diamond ring. As soon as you see the photosphere, look away from the Sun and check once again for shadow bands. They could be very strong immediately after totality. As the bands fade, within seconds or minutes, the Sun quickly develops into a thicker crescent and, after another hour or so, the eclipse is over. And while the physical event is over, the memories of it last the rest of your life.

ENDNOTES

[1] F. Espenak, M. Littmann, and K. Willcox, *Totality: Eclipses of the Sun,* 2nd edn. New York: Oxford University Press, 1999, 233.
[2] S. Koutchmy, F. Baudin, K. Bocchialini, J.-P. Delaboudinière, and A. Adjabshirizadeh, The August 11th, 1999 total eclipse CME. *Proc. SOLSP: The Second Solar Cycle and Space Weather Euroconference,* ESS SP-477, 2002, 55–58.

# 10

## Solar eclipse photography

Then here I take my leave of thee, fair son,
Born to eclipse thy life this afternoon,
Come, side by side together live and die,
And soul with soul from France to heaven fly.

(Shakespeare, *I Henry VI*, 4.5.52–55)

If *seeing* a total eclipse of the Sun is great, then actually *photographing* one must be better. Right? Not necessarily. On many total eclipses, I have advised first-time viewers not to try photography. The experience is so emotional, so engrossing, and so brief, that it would seem a waste of precious time to worry about taking pictures, especially since, if you wait a period of time, you will see professional pictures of the eclipse. It is logical advice, but almost no-one takes it. First eclipse, take a picture; that is what most people do.

So what camera should you use? What settings? If you are using film, what film is best? We begin with 10 simple rules, and then proceed by way of questions and answers based on the many queries I have heard from eclipse watchers anxious to make their first eclipse photography a successful and pleasant one.

### Ten "commandments" for total solar eclipse chasers

(1)   If this is your first eclipse, please do not try to photograph the total phase. Just concentrate on enjoying it. But since almost everyone ignores this advice . . .

(2)   Never ever *ever* try to photograph the partial phases of a solar eclipse without filters or projection of the Sun's image. You can damage your retinas, and burn out your camera.

**Figure 10.1** With an automated digital camera, it is hardly possible, or at least extremely difficult, to ruin an eclipse shot. In this picture with a Canon PowerShot 5-megapixel "point-and-shoot" camera, the eclipsed Sun is shown clearly with its bright, near-circular corona typical of eclipses which occur near the minimum of a solar cycle. Mercury and Venus are within the field of the camera, however, but the automated settings that concentrated on the Sun did not record them. Photograph by Wendee Wallach-Levy.

(3)  It's hard to take a bad picture of the total phase of a solar eclipse. Setting a digital camera on automatic and snapping away should produce acceptable results.

(4)  Plan your photographic mission carefully and rehearse before totality. Try photographing the Moon at night to get a sense of how it will appear through different lens combinations. The total phase will go by much more quickly than you can imagine. (Remember California amateur astronomer Norm Sperling's seven second rule: from the point of view of experiencing one, all total solar eclipses last seven seconds.)

(5)  Bracket your exposures. No matter how much you read about setting exposure times, during totality it is best to take a number of pictures at different settings by changing either the aperture or exposure settings. From 1/125th second to 5 seconds is a good

range. This way, you will get a range of images that show both the faint detail close to the Sun, as well as the full range of detail in the corona.

(6)    Don't forget the landscape or seascape. The onrushing shadow just before totality, during totality, and especially as it races away from you after totality, offer excellent photographic opportunities.

(7)    If your camera malfunctions even slightly, don't waste precious seconds trying to repair it. Just enjoy the eclipse.

(8)    Before totality, make sure that your flash is turned off and will not activate automatically during totality; a flash will not improve your picture and will bother those around you who are trying also to observe the eclipse.

(9)    Try to take pictures of the crescent Sun projected between leaves, or small holes cut in paper.

(10)    Don't forget to enjoy the eclipse!

## Basic solar eclipse photography

*Camera and lens*

Any camera can take a good eclipse picture. The cheapest disposable camera can show an image of a crescent Sun projected onto a sheet of paper or the wall of a building. During totality, the wide angle view of a simple camera captures the broad swath of the Moon's shadow as it passes over you, its sharp edge allowing light to creep in around the horizon. Some of the most dramatic pictures I have seen come from these cameras.

With more advanced cameras, you can capture detailed images of the crescent Sun, the corona around the Sun, and the landscape or seascape around you.

## Should I use film or should I use a digital camera?

With the extraordinary development of digital cameras in recent years, the idea of taking solar eclipse pictures directly onto a chip, rather than by using film, has taken hold. Eclipse watchers today overwhelmingly use digital cameras, partly because they are so readily available and partly because they are so much easier to use than the older film cameras.

It is interesting to remember that with the advent of digital cameras and computer use in photography, it would seem that a century of learning on exposure settings and depth of field has gone out the window. That isn't true at all. The basic photography information that you and famous photographic artists like Ansel Adams both need in order to get that perfect picture, has not changed substantially. The *only* change is in the medium used to record the image. Instead of a chemical emulsion that is moderately sensitive to light, digital cameras use a charge-coupled device that is *extremely* sensitive to light. If the camera is automated, then it will set the aperture and exposure aperture for you. But in more advanced digital cameras, you can set the ISO (International Standards Organization) number which represents the film speed. You can set the aperture of the lens either to use most of it, and get a narrow range of focus, or less of it to get a greater depth of focus. You can also set the shutter speed. Cameras let you go halfway these days – you can set the lens at a specific focal or f/ratio, and the camera will automatically calculate the correct exposure length. Conversely, you can set the shutter speed, say at 1/125 second, and the camera will calculate the correct f/stop. With these choices, eclipse photography becomes even more fun – and more complex.

With film, you could run out of your roll in the middle of a total eclipse with neither time nor inclination to change rolls. But the same problem could happen with the memory chip of a digital camera should it become filled at the wrong time. It is important to be aware of these limitations.

### What lens should I use?

The picture you get will depend largely on your choice of camera and lens. While a beginner-level automatic camera will produce pleasing images, you will want a longer lens, or even a telescope. During the 1979 total eclipse, I used a Celestron C-90 telescope (really a 3.5-inch diameter, f/11 lens). It had a 900-mm lens and produced a very pleasing image size of the Sun. Typically, a 500–2000-mm lens or telescope will work well. With the 2000-mm, the Sun's image will almost fill the field of a typical modern camera. To determine the size of the Sun's image on whatever system you use, divide the focal length of the lens by 110.

### How long should the exposures be?

During the two minutes of totality in 1979, I set the aperture stop at f/8 and ran through the entire range of camera exposure settings: from

1/500 second to about 5 seconds. I had hoped that with that range, one shot might work. I went through the range several times as the period of totality progressed. The great surprise: almost every one of the pictures was successful! Although 1/500 second was too fast, even that one recorded the details of the prominences that appeared during the second half of totality. The shorter exposures recorded prominences and various amounts of the inner corona, and the longer ones displayed the grand spectacle of the outer corona as it shot out from the Sun. Because I went through the range in cycles, even exposures of the same duration, taken at different points during the total phase, revealed changing details. The first short exposure of my first cycle did not record any prominences, but the short one at the start of the second cycle did, as the Moon's slow advance across the Sun permitted them to come into view.

I used Ektachrome 200 film during this eclipse. This medium-speed film was not too grainy, and certainly sensitive enough for exposures of the Sun. Now I use a digital camera instead. The speeds and settings can emulate those of the film. During the 2006 eclipse, I decided to set the ISO speed at 200 and simply run through the shutter speed settings just as I did in 1979. Within a few seconds of the start of totality, however, I knew that my plan wasn't going to work. Why? Beyond the edge of the Moon's shadow are light areas in the sky near the horizon that let enough light into the area of totality that it is easy to read camera settings. This time, low clouds seemed to hang around the horizons – we were incredibly lucky that the temperature drop in the area of totality was enough to make the clouds above us disappear. But the horizon clouds prevented the surrounding light from brightening up where we were, making this eclipse the darkest one I had seen since 1970, and that one was clouded out. It was so dark that I couldn't even read my camera settings. I had no choice but to set everything on automatic and accept whatever I got!

Keep careful notes on the particular settings for each picture you take, so that you can later know which settings you are happiest with. This advice is much easier given than complied with. Spending time writing down numbers on a form wastes precious seconds. But you can talk your way through the sequence, with a recorder to record your voice. Or, prepare a plan ahead of time and follow it strictly during the moment of totality.

### How do I photograph the partial phases?

The best way to capture the partially eclipsed Sun is to use a camera with a lens large enough so that the Sun will occupy much of the field. So a lens from 500 mm or larger will be best.

You need to use a filter to cover your camera. See Chapter 4 for a discussion about filters.

Run a rehearsal with your equipment some time before the eclipse, and certainly well before you travel to see the eclipse. Start with an automated exposure, and then bracket with different exposure speeds and shutter settings.

### How does a photographer reduce vibration?

The best way is to make sure the camera is well supported. Mounting the camera on a tripod and using a cable release helps ensure that the images will be steady because there is no direct physical contact with the camera during the exposure. The cable is an ingenious solution to the problem of an unsteady hand touching the camera at the critical moment. With a cable, the only contact you have is with the end of the cable, thereby keeping the camera steady.

### Can I use projection?

Try projecting the Sun using a telescope onto a screen. There is a big caveat, however. A traditional Newtonian reflector, or refractor with star diagonal, would work. I do not recommend using Schmidt–Cassegrains without filters for projecting the Sun. The complex baffling system inside can heat up and catch fire.

The real problem with photographing a projected image is that it really takes three people to manage it – one for the telescope, another for the cardboard projection screen, and a third to take the picture.

### Don't forget to focus!

Just set the focus at infinity, right? Indeed, this procedure should work, but especially when long lenses are used, it is important to focus

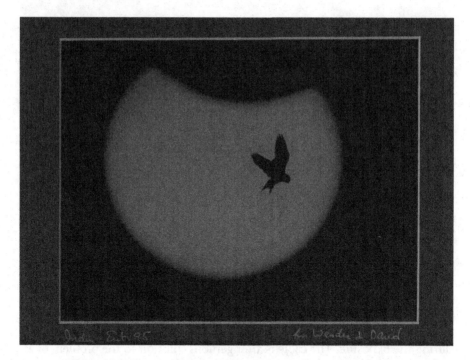

**Figure 10.2** No matter how much experience a photographer might have, being ready for and taking advantage of the unexpected is what spells the difference between a good photographer and a great one. In this photograph, Eliot Schechter captured this gull lazily crossing in front of the eclipsed Sun during the partial phases of the annular eclipse of April 29, 1995.

anyway, using a magnifying lens if one is available on your camera. Focus ahead of time, then tape over the focusing ring on your lens, or the focus knob on your telescope, to prevent any unplanned and unwelcome change.

## Videography

If still photos are good, how about videos? During the 2005 eclipse, totality lasted only 29 seconds. Wendee decided to try taking a video of the eclipse. The movie came out beautifully. The experience was satisfying because she was able to *see* the eclipse at the same time, by holding her camera about 10-inches away and using the viewscreen to center the eclipse. By looking over the camera she was also able to witness the event she had traveled so far to see.

Saros 124

A SOLAR SAROS CYCLE

| Julian Calendar | | | | | |
|---|---|---|---|---|---|
| 1049—Mar 6 p | 1445—Oct 30 t | 1842—Jul 8 t |
| 1067—Mar 17 p | 1463—Nov 11 t | 1860—Jul 18 t |
| 1085—Mar 28 p | 1481—Nov 21 t | 1878—Jul 29 t |
| 1103—Apr 8 p | 1499—Dec 2 t | 1896—Aug 9 t |
| 1121—Apr 18 p | 1517—Dec 13 t | 1914—Aug 21 t |
| 1139—Apr 30 p | 1535—Dec 24 t | 1932—Aug 31 t |
| 1157—May 10 p | 1554—Jan 3 t | 1950—Sep 12 t |
| 1175—May 21 p | 1572—Jan 15 t | 1968—Sep 22 t |
| 1193—Jun 1 p | | 1986—Oct 3 t |
| | Gregorian Calendar | |
| | 1590—Feb 4 t | 2004—Oct 14 p |
| 1211—Jun 12 t | 1608—Feb 16 t | 2022—Oct 25 p |
| 1229—Jun 22 t | 1626—Feb 26 t | 2040—Nov 4 p |
| 1247—Jul 4 t | 1644—Mar 8 t | 2058—Nov 16 p |
| 1265—Jul 14 t | 1662—Mar 20 t | 2076—Nov 26 p |
| 1283—Jul 25 t | 1680—Mar 30 t | 2094—Dec 7 p |
| 1301—Aug 5 t | 1698—Apr 10 t | 2112—Dec 19 p |
| 1319—Aug 16 t | 1716—Apr 22 t | 2130—Dec 30 p |
| 1337—Aug 26 t | 1734—May 3 t | 2149—Jan 9 p* |
| 1355—Sep 7 t | 1752—May 13 t | |
| 1373—Sep 17 t | 1770—May 25 t | |
| 1391—Sep 28 t | 1788—Jun 4 t | |
| 1409—Oct 9 t | 1806—Jun 16 t | |
| 1427—Oct 20 t | 1824—Jun 26 t | |

p = partial eclipse
t = total eclipse

**Figure 10.3** A very odd photograph of a partial solar eclipse. I purchased this book, *A Key to the Heavens*, on July 31, 1964 while at the Adirondack Science Camp when I realized that (a) it was a book about astronomy, which was good enough for me, and (b) it contained a table of the saros series that happened to include the October 3, 1986 annular eclipse that I witnessed as a slight partial. It also listed the following eclipse in the same saros 124, a partial on October 14, 2004. This picture was captured during that eclipse, near sunset atop Mauna Kea, Hawaii. It includes the shadow of Cupid, the Questar telescope I used that day, and visible within two pinholes of that shadow are images of the eclipse in progress (the lower right one is the superior one, clearly showing a crescent Sun).

Video cameras use either CCD (charge-coupled devices) or MOS (metal-oxide semiconductors). They don't seem to be bothered if they are pointed briefly at the Sun. Most cameras have zoom lenses. Don't pay any attention to what the camera says about high "digital zoom" – that is a zoom feature that reduces the resolution of the image substantially. "Optical zoom" is what you are interested in, since that value physically changes the image scale. As I suggested in the section on still photography, experiment by photographing the Moon at the highest zoom. To capture the grandeur of the corona during total eclipse, you will need at least twice as much field as that.

**Table 10.1.** *The size of the Sun's image for various photographic focal lengths*

| Focal Length | Field of View (35 mm) | Field of View (digital) | Size of Sun |
|---|---|---|---|
| Field of View and Size of Sun's Image for Various Camera Focal Lengths | | | |
| 14 mm | 98° × 147° | 65° × 98° | 0.2 mm |
| 20 mm | 69° × 103° | 46° × 69° | 0.2 mm |
| 28 mm | 49° × 74° | 33° × 49° | 0.2 mm |
| 35 mm | 39° × 59° | 26° × 39° | 0.3 mm |
| 50 mm | 27° × 40° | 18° × 28° | 0.5 mm |
| 105 mm | 13° × 19° | 9° × 13° | 1.0 mm |
| 200 mm | 7° × 10° | 5° × 7° | 1.8 mm |
| 400 mm | 3.4° × 5.1° | 2.3° × 3.4° | 3.7 mm |
| 500 mm | 2.7° × 4.1° | 1.8° × 2.8° | 4.6 mm |
| 1000 mm | 1.4° × 2.1° | 0.9° × 1.4° | 9.2 mm |
| 1500 mm | 0.9° × 1.4° | 0.6° × 0.9° | 13.8 mm |
| 2000 mm | 0.7° × 1.0° | 0.5° × 0.7° | 18.4 mm |

*Note:* Size of Sun's Image (mm) = Focal Length (mm)/109.
www.mreclipse.com/SEphoto/SEphoto.html © *2008 Fred Espenak.*

**Table 10.2.** *Solar eclipse exposure guide. These tables are reproduced with the kind permission of Fred Espenak, cf. http://www.mreclipse.com/LEphoto/image/LE-Exposure1w. GIFGuide*

| fl | 32 | 64 | 100 | 200 | 400 |
|---|---|---|---|---|---|
| **SUN – full disk or partial eclipse**; Through full aperture Solar Filter Film Speed – ISO | | | | | |
| **2.8** | 1/1000 | 1/2000 | 1/4000 | – | – |
| **4** | 1/500 | 1/1000 | 1/2000 | 1/4000 | – |
| **5.6** | 1/250 | 1/500 | 1/1000 | 1/2000 | 1/4000 |
| **8** | 1/125 | 1/250 | 1/500 | 1/1000 | 1/2000 |
| **11** | 1/60 | 1/125 | 1/250 | 1/500 | 1/1000 |
| **16** | 1/30 | 1/60 | 1/125 | 1/250 | 1/500 |
| **22** | 1/15 | 1/30 | 1/60 | 1/125 | 1/250 |
| **32** | 1/8 | 1/15 | 1/30 | 1/60 | 1/125 |
| **SUN – total eclipse: prominences**; No filter Film Speed – ISO | | | | | |
| **2.8** | 1/500 | 1/1000 | 1/2000 | 1/4000 | – |
| **4** | 1/250 | 1/500 | 1/1000 | 1/2000 | 1/4000 |
| **5.6** | 1/125 | 1/250 | 1/500 | 1/1000 | 1/2000 |
| **8** | 1/60 | 1/125 | 1/250 | 1/500 | 1/1000 |
| **11** | 1/30 | 1/60 | 1/125 | 1/250 | 1/500 |
| **16** | 1/15 | 1/30 | 1/60 | 1/125 | 1/250 |
| **22** | 1/8 | 1/15 | 1/30 | 1/60 | 1/125 |
| **32** | 1/4 | 1/8 | 1/15 | 1/30 | 1/60 |

**Table 10.2.** (*cont.*)

| fl | 32 | 64 | 100 | 200 | 400 |
|---|---|---|---|---|---|
| **SUN – total eclipse: inner corona (3\* field)**; No filter Film Speed-ISO | | | | | |
| **2.8** | 1/125 | 1/250 | 1/500 | 1/1000 | 1/2000 |
| **4** | 1/60 | 1/125 | 1/250 | 1/500 | 1/1000 |
| **5.6** | 1/30 | 1/60 | 1/125 | 1/250 | 1/500 |
| **8** | 1/15 | 1/30 | 1/60 | 1/125 | 1/250 |
| **11** | 1/8 | 1/15 | 1/30 | 1/60 | 1/125 |
| **16** | 1/4 | 1/8 | 1/15 | 1/30 | 1/60 |
| **22** | 1/2 | 1/4 | 1/8 | 1/15 | 1/30 |
| **32** | 1 sec | 1/2 | 1/4 | 1/8 | 1/15 |
| | | | | | |
| **SUN – total eclipse: outer corona (10\* field)**; No filter Film Speed-ISO | | | | | |
| **2.8** | 1/2 | 1/4 | 1/8 | 1/15 | 1/30 |
| **4** | 1 sec | 1/2 | 1/4 | 1/8 | 1/15 |
| **5.6** | 2 sec | 1 sec | 1/2 | 1/4 | 1/8 |
| **8** | 4 sec | 2 sec | 1 sec | 1/2 | 1/4 |
| **11** | 8 sec | 4 sec | 2 sec | 1 sec | 1/2 |
| **16** | 15 sec | 8 sec | 4 sec | 2 sec | 1 sec |
| **22** | 30 sec | 15 sec | 8 sec | 4 sec | 2 sec |
| **32** | 60 sec | 30 sec | 15 sec | 8 sec | 4 sec |

*Note:* © Copyright 1999 by Mark Littmann, Ken Willcox, and Fred Espenak.

Since the partial phases add more than two hours to the full duration of the eclipse, you can try taking short exposures lasting a couple of seconds every five minutes. This will "rush" the long partial phases into just a few minutes. That leaves the short time span that begins as the crescent Sun shrinks to a line of light, then breaks up into Bailey's beads, then disappears entirely in the flash of the diamond ring. For the next few minutes the world has changed as the corona and prominences rule the sky. No matter how good your movie or stills will be, nothing – absolutely nothing – can replace the raw emotion and thrill of looking up and taking in the whole experience of a total solar eclipse.

### Exposure table

Reproduced by the gracious permission of Fred Espenak, Table 10.2 suggests exposures for eclipses. But as I am sure he will also admit, this table is only a guide. A light cover of cloud will still allow you to enjoy the eclipse but wreak havoc with all your pictures. Again, my best advice is to bracket your exposures.

# Part III   ECLIPSES OF THE MOON

## 11

# Don't forget penumbral lunar eclipses!

Crooked eclipses 'gainst his glory fight . . .

(Shakespeare, *Sonnet 60.7*)

It is possible that the single most delightful observing session can happen when there is a penumbral eclipse of the Moon. Such an event is as far removed from a total solar eclipse as one can get, and still have an eclipse of some sort. In a penumbral lunar eclipse, the full moon enters, travels through, and then exits the outer shadow of the Earth. This part of the shadow is called the penumbra, and its effects can range from absolutely nothing to, at best, a slight darkening of one edge, or limb, of the Moon. For almost all observers, that is all. For Thomas Hardy, whose wonderful poem "At a Lunar Eclipse" appears at the start of Chapter 13, the total eclipse he saw in 1903 was preceded by a penumbral shading.

However, for me there is much more. At no other time do the rays surrounding the younger craters, like Tycho and Copernicus, appear so obviously. These big craters are certainly not young by human standards; Tycho was formed about 100 million years ago by the sudden impact on the Moon of a comet or an asteroid, and Copernicus is a bit older, but both are young geologically. The Moon's oldest craters were formed probably by impacts mostly during a period about 3.9 *billion* years ago, and is remembered today as the period of late heavy bombardment. As Carolyn Shoemaker said, "The inner solar system was bombarded twice, once during its formation, and again when it got the stuffing knocked out of it." Later craters like Tycho were formed when more recent objects struck the Moon; in fact there is a theory that holds that the object (a comet or an asteroid) that formed Tycho might have been related to the one that later formed a giant crater near the present town Chixulub, on Mexico's Yucatán peninsula, the result of an impact that probably led to the demise of more than 75% of all species of life on Earth, including the dinosaurs. It is possible that the source of this brief increase in inner solar system bombardments was the catastrophic breakup of the parent body of what is now the asteroid Baptistina, about 160 million years ago.[1] In the case of the impact that created Tycho, at the moment of collision, large amounts of debris took off and landed again in other spots, so that Tycho's ray system covers almost half the surface of the Moon. The rays always come across better at full Moon than at any other time, but if the

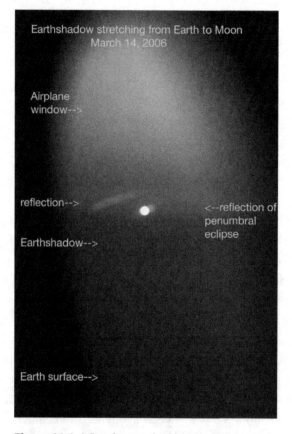

**Figure 11.1** A "total penumbral lunar eclipse" seen from an aircraft window. This photograph captures the shadow of the Earth rising out of the eastern sky and extending as far as the Moon. The best view of the eclipse is actually the reflection of the main image on the glass of the airplane window. Photograph by David H. Levy using a Canon 20 DA camera optimized for astronomical imaging.

**Figure 11.2** A penumbral eclipse of the Moon imaged through passing clouds using Obadiah, my Schmidt camera. The presence of the cloud allowed the photograph to happen, otherwise the Moon would have been far too bright to image with a Schmidt camera. Photograph by David H. Levy.

Moon's phase is slightly darkened, as during a penumbral eclipse, the rays become more prominent.

At what point during the Moon's entry into the Earth's shadow does an eclipse become visible? The standard answer is at least 50 percent. If the Moon enters that much of the shadow, it is likely that the eclipse will be noticeable. But what about 40 percent? Or 35?

The answer depends partly on the strength of the Earth's shadow, which is a direct result of the amount of dust in the atmosphere. A volcanic eruption that spews particles of dust into the atmosphere will darken the shadow considerably, thus allowing a shallower penumbral eclipse to be more easily visible, whereas under normal conditions the Moon would show no darkening whatsoever.

So it was with some excitement that on April 24, 2006, just two weeks after a lovely total eclipse Wendee and I enjoyed from a ship sailing the Aegean Sea,

I set out to observe a penumbral lunar eclipse. The evening before, I was locked under clouds and rain. The clear sky chart[2] showed the possibility of a hole in the clouds to the west, so I thought I might have to do a bit of predawn travel to catch a glimpse of this one. As Wendee and I went to bed late that evening, our power failed, came back on, then failed again. When it came back on once more a half-hour later, I quickly opened the garage's electric door and moved our car outdoors in the rain to avoid another failure preventing me from taking the car out later that night, when the eclipse was scheduled to occur.

When my alarm went off at 2:25 am I looked outside to dense overcast but at least it had stopped raining. Outdoors, I took stock of the opaque sky. The cloud deck was moving from the southwest, and I persuaded myself that low toward that horizon the sky seemed just a bit brighter. At 2:40, with just 10 minutes to maximum eclipse, I loaded Echo, the 3.5-inch f/11 Skyscope that I had used since the summer of 1960, into the car and headed west.

If I had any hope of catching the maximum phase, I was wise to do that. The further west I drove the more obvious that break was looking. This late in the night, my first telescope and I had the road pretty much to ourselves. Twenty miles from home, I reached a corner where a traffic signal protected a large but completely deserted intersection. I turned left, pulled over, took Echo out of the car, and waited a few minutes for the Moon to appear. My plan was to hand-hold the telescope just in case I needed to make a quick getaway from this instant roadside observatory.

Even through light clouds, I could clearly see the Moon's darkened northern limb, its mountains projecting sharply against the sky. I guessed that had this been a total eclipse, it would have been a bright one, with a Danjon rating of 4 or 5, typical of a relatively dust-free atmosphere of the Earth. But it is hard to tell with so little shadow to go by. The rays of the younger lunar features were not as pronounced as they usually are during penumbral events, but they were clearly visible. I was feeling so good, having beaten the odds to catch the maximum, that I didn't notice the approach of another car. I turned around to see a police cruiser right behind me, its red lights flashing and its occupant nervously eyeing the long black object I was holding.

I walked over, introduced myself, and told the officer that I was holding a telescope, and that I was using it to see the eclipse. I asked him if he thought that the Moon really was full. Right away he looked at the Moon and told me about its darkened northern limb. We shared and enjoyed a successful teaching moment. After a brief conversation about eclipses, he bid me good morning and left. I love these unexpected teaching opportunities. With the onset of more clouds, I headed home, but another clearing stopped me in front of a beautiful grove of giant saguaro cactus plants that marked the site where I had gotten some pretty neat pictures of brilliant Comet Hyakutake a decade earlier. The Moon had begun to swing away from the Earth's shadow, but I could still clearly see it.

Finally I was home, my telescope safely set up in our backyard. It was now one hour past mid-eclipse, and only about a third of the Moon still was within the shadow. Even with its unusual lightness, I could still make out its presence on the Moon's northwest side. I had never detected a penumbral eclipse with that little of the shadow on the Moon. Usually 50% is the cutoff point between a penumbral eclipse that can be observed and one that cannot. Ten minutes later, I could no longer detect any sign of the shadow.

So ended my penumbral eclipse adventure. It was neither as long as the solar eclipse equivalent Wendee and I had just concluded near Greece, nor was the eclipse as dramatic. But, according to Fred Espenak's Eclipse Home Page,[3] it was the 23rd of a series of eclipses that began as an undetectable penumbral event on August 25, 1608, during Shakespeare's last years as a playwright, as saros 141. Penumbral after penumbral followed; I saw the previous one on April 14, 1987. The first partial will occur on May 16, 2041, the first total on August 1, 2167. This particular saros fades away with an undetectable penumbral on October 23, 2906.

### A plane eclipse

Both astronomy and our modern world deal with flight. The Sun, the Moon, and Earth fly through space, their control systems, following Isaac Newton's laws, determining where and when they will all line up and produce an eclipse. The Newtonian world is so precise that the penumbral lunar eclipse of March 14, 2006 could have been predicted 100,000 years ago. But the airline flights that get us to these eclipses are part of our own less predictable modern world, and so the margin for error is high until the plane lifts off the ground.

On March 14, 2006, a rare "total penumbral" lunar eclipse took place. An observer standing *anywhere* on the Moon's surface would enjoy a partial eclipse of the Sun. From Earth, however, the eclipse would be over by the time the Moon rose over our home in Vail, Arizona. I knew that I could see it only if I could just travel 1000 miles to the east. I chose American Flight 2606 to Chicago, and arranged for seat 21F on the MD80 so that I would have a roomy exit row view to the south. Since the plane would be racing to the northeast, I hoped for a view of moonrise.

From 36,000 feet the sky is a deep, contrasty blue. The Moon rose before sunset, its upper right quadrant significantly darkened. But what happened next was a surprise. Minutes after sunset the Earth's great shadow soared out of the eastern horizon, its long cone reaching all the way to the Moon in eclipse. From tiny window 21F I could see the magnificent art of Earth's dark shadow touching the Moon.

To conclude, I love eclipses – and that enthusiasm includes all the events that involve the Sun, Earth, and Moon lining up. Those events run the gamut between total eclipses of the Sun and penumbral eclipses of the Moon. Why bother to observe those things where the Moon fails even to enter the Earth's central shadow? The best you can see during a penumbral lunar eclipse is a slight shading on the side of the Moon closest to the umbra. But for me, these events are great: as the Moon travels eastward, the Earth's shadow shades the Moon's northeast or southeast rim (these directions are the sky, not the Moon), and it's fun to watch the faint shadow slowly slide across the limb, causing features like mountains to appear in stark contrast to the more-or-less featureless rest of the full moon.

ENDNOTES

[1] W. F. Bottke, D. Vokrouhlicky, and D. Nesvorny, An asteroid breakup 160 Myr ago as the probable source of the K/T impactor, *Nature* 449 2007, 48–53.

[2] http://cleardarksky.com/c/ JarnacObAZkey.html

[3] http://eclipse.gsfc.nasa.gov/eclipse.html

# 12

## Partial lunar eclipses

The mortal moon hath her eclipse endured . . .

(Shakespeare, *Sonnet 107.5*)

### Similitudes

Watching a lunar eclipse is like many things. It is like reading *Hamlet* and understanding every word, every cadence. It is like having a candlelit dinner at home. It is like watching your favorite movie. But even more suggestive is what a lunar eclipse is *not* like. It is not like a total solar eclipse, with its massive emotional buildup, brief exultation at totality, then a partial letdown after. An eclipse of the Moon is more like a two-hour long massage, pure relaxation followed by joy at the event's having taken place over your head. So long as the weather is clear, you really can't go wrong with a lunar eclipse.

In the spring of 1997, as Wendee and I were preparing for our March 15 wedding, we faced the problem of our on-campus ceremony being marred or postponed by crowds attending a basketball NCAA playoff game. We telephoned Mike Terenzoni at the Flandrau Science Center, the place we had chosen for our nuptials. "What about changing it to one week later, March 23?" we asked. "Not then," his answer came. That's Hale–Bopp."

I considered the impact of having a wedding on a night near closest approach to the Earth of the great comet of 1997. "Actually," I explained, "Hale–Bopp will be bright for so long that it really doesn't matter." But Michael was already looking at his calendar. "And," he added, "there's a lunar eclipse that night."

**Figure 12.1** A view of a partial lunar eclipse. Photograph by David H. Levy.

The thought of an eclipse of the Moon happening on our wedding night was just too good to resist. Notwithstanding the idea that starting a marriage with an eclipse is a good, or not a good, omen, we thought it would be fun to try. After the afternoon ceremony, luncheon, dancing, and a planetarium show, we assembled at our home to watch the eclipse. We beat a dire forecast for cloudy weather; by nightfall the sky was clear except for some cloudiness to the east, above which the Moon rose well before the umbral phase of the eclipse began. With the eastern edge of the Moon approaching the umbra, about 10 minutes before first umbral contact, the Moon began to darken. In the next few minutes the Earth's shadow began silently to cross the Moon like a large broom. Although this was not to be a total eclipse, the effect was similar as more than 90 percent of the Moon was immersed in shadow at maximum eclipse. This was enough to darken the sky so that, despite the fact that it was a night of full moon, the sky became dark enough to allow a good deal of observing of faint deep sky objects like the faint galaxies huddled within Leo, Coma Berenices, and Virgo. This place marks the center of the "Local Supercluster" of galaxies of which our own Milky Way is a member. It was a superb evening. We had set up a place on the south side of our border fence so that serious amateur astronomers could observe and photograph the eclipse without interruption from the other party guests. The plan worked very well, except for some occasional remarks about "the guys on the other side of the tracks," who seemed to be there more for the eclipse than to wish us a happy marriage! These people were viewing and photographing the progress of the eclipse, and also enjoyed rare sightings of faint deep sky objects on a full moon evening. This is one important aspect of what a lunar eclipse is about, since it offers a chance for viewers to enjoy objects not normally visible when a bright moon is in the sky.

### Timing crater contacts

Watching the Earth's shadow cross the rims of craters on the Moon is one of the truly exciting things to do during an eclipse. In the cold recording of hour, minute, and second, we sense the wondrous motion of Earth and Moon through space. The soft, inexorable motion of the shadow testifies to that slow but unstoppable moment.

Timing crater shadow contacts was one thing I was looking forward to doing on the evening of Wednesday, June 24, 1964. The day had been mostly clear and I was excited about viewing the lunar eclipse that would begin at dusk. At dinner, however, the sky began to darken quickly."It looks like rain," Dad innocently remarked. The argument that ensued was brief and unpleasant. The clouds did cost me the total phase of the eclipse, but the sky began to clear in time for the closing of the partial phase. My view that evening was limited to seeing a bright Moon with its top lopped off, faded to darkness by the Earth's shadow. For me, the great lunar eclipse of June 1964 was but a shallow partial eclipse.

What did I learn that night? First, that even though Dad was not always right, he was never wrong. Although that night was an exception, he and I usually got along pretty well. We made up a few days later, but I never got to share that eclipse with him.

### The Minnowbrook partial eclipse of 1970

During the late summer of 1970 I managed to turn both a camp movie and the outbreak of the Olympic Games at Camp Minnowbrook into educational venues for a lunar eclipse. Of all the times I have worked at various summer camps over the years, on only this occasion did an eclipse of the Moon liven up my summer observing nights. Built on the site of one of the old Adirondack great camps, this camp was designed to bring the arts and sciences into a camping experience.

Clouds hung around most that day, and they persisted into the early evening, threatening our attempt to observe the eclipse. Lothar Eppstein, the camp director, approved an "astronomy watch" for the eclipse conditionally but would not allow us to begin observations until after the movie ended. That particular movie, *The Solid Gold Cadillac*, was a yarn about a good man trying to succeed in his business despite having to work within a corrupt group. After the first reel I asked if I could take some of the more advanced kids and set things up. Lothar denied my request but did allow me to drift off occasionally to see the start of the eclipse under a sky that, miraculously, was starting to clear. But after the movie came the surprise announcement that the annual Camp Minnowbrook Olympics had started. Teams were then announced after the movie; Lothar divided the camp into two teams, Athenians and Spartans. I was to be on the Athenian side. Suddenly Lothar winked at me and commanded something like, "Now Levy and his group – get out there and enjoy the lunar eclipse!" We were on. We went outside at about 10:30 pm, set up the several telescopes, and began waiting for the Moon to emerge from behind a cloud. Fortunately it complied, and by the time the other campers had joined us, the eclipse was under way, less than an hour before maximum.

Levy's group eventually swelled to include about half the camp, campers and staff included. Organizing a program that combined semi-scientific observations with astronomy education was a challenge, but I did get some children to time crater contacts, while others recorded color and brightness changes as the Earth shadow marched across the Moon. At 11:30 that evening, about half the Moon was in eclipse. Also, around mid-eclipse clouds began to gather once more, rendering the second part of the eclipse visible only intermittently.

Despite these challenges, this partial lunar eclipse was one of my favorites. It turned out to be one of the most successful teaching opportunities of all the years I spent at any summer camp. The children learned several things, not the least of which was that natural events occur at specific times; we cannot look up at any hour of the day or night and see an eclipse. But when an eclipse does come along, we should be ready for it. Further, we can learn that this eclipse, like every eclipse, has a lineage – a history and a future. It was preceded by one of slightly greater magnitude – meaning that more of the Moon was immersed in the Earth's shadow, precisely 6,585 1/3 days earlier, and another will follow by that same span of time, to occur a little more than 18 years in the future. The earlier eclipse took place on August 5, 1952, long before any of the 1970 campers were born. Historically, 1952 was near the height of the McCarthy anti-communist rampage in the United States and, except for a few geologists, the Moon was just a pretty object in the sky; by 1970 four Americans had walked on its surface. This saros, it turns out, is fading; its last total eclipse took place on June 22, 1880, long before automobiles, television, telephones, or for that matter anything that was powered by electricity.

Late as it is, this "Minnowbrook saros" has a long way to go. I saw the one that followed, a lesser partial eclipse on August 27, 1988. While the 1970 eclipse was an evening event in the northeast part of North America, the 1988 eclipse was visible as a predawn happening on the west coast, and there was one smaller still on September 7, 2006. There will be a miniscule one on September 18, 2024. Thus, an eclipse brings with it a history lesson, plus a question about humanity's future. In 1880 the United States was still recovering from its Civil War, and the previous total one, in 1862, took place during one of the darkest periods of that war, although it was invisible from the United States. Quite likely no one in that 1970 summer group could have forecasted how personal computers would make inroads by 1988, and that these devices, small enough to sit atop our laps or even in our pockets, would order our lives long before 2006.

Besides crater contact timings and color change observations, the main thing to do during a partial lunar eclipse is to enjoy it. An eclipse is a reason to pause and reflect upon a moment in time, a chance to see Nature at its best.

# 13

## Observing a total eclipse of the Moon

Thy shadow, Earth, from Pole to Central Sea,
Now steals along upon the Moon's meek shine
In even monochrome and curving line
Of imperturbable serenity.

How shall I like such sun-cast symmetry
With the torn troubled form I know as thine,
That profile, placid as a brow divine,
With continents of moil and misery?

And can immense mortality but throw
So small a shade, and Heaven's high human scheme
Be hemmed within the coasts yon arc implies?

Is such a stellar gauge of earthly show,
Nation at war with nation, brains that teem,
Heroes, and women fairer than the skies?
(Thomas Hardy, "At a Lunar Eclipse," 1903)

Of the many eclipses of the Moon that I have enjoyed, none can approach, or will likely ever come within reach of the unbelievable blackness of the eclipse of December 30, 1963. That eclipse had the Moon pushing its way through an atmosphere clouded with dust from a recent volcanic eruption. Constantine Papacosmas, an experienced observer who saw it with friends, barely saw the Moon at all, and then only faintly through binoculars. He estimated that the eclipsed Moon was no brighter than a fifth-magnitude star. From where I was, finding the Moon in a city sky was almost impossible. I recall enjoying the clear, bitterly cold night, then rushing indoors to sit atop an electric heater.

## The Danjon scale for measuring brightness of lunar eclipses

Just how dark was that eclipse? To measure its luminosity we use a scale that ranges from 0 to 4. Proposed by André-Louis Danjon during the first half of the twentieth century, this visual scale effectively measures the visibility of the Moon during the middle part of a total eclipse. Although the scale is effective for eclipses, it was not developed for that purpose at first. Danjon created a prismatic system in which a prism inserted into a telescope split the view of the Moon into two discrete images. Danjon could dim one of the images using a diaphragm until the sunlit part of the Moon was darkened to the same apparent brightness as the Moon's unlit portion (or more correctly, the portion of the Moon lit by reflected light from the Earth, or earthshine). In this way Danjon could record the brightness of earthshine. Adjusted for lunar eclipses, the values, recorded as *L*, are as follows:

**Figure 13.1** Messier 101. Photographed through Clyde, my 14-inch Celestron Schmidt–Cassegrain telescope (SCT) with hyperstar lens attached to the front lens. This image was actually taken remotely, from a hotel not far from Twin Valleys Camp, where our annual Adirondack Astronomy Retreat is held. The image contains stars as faint as nineteenth magnitude. Photograph by David H. Levy.

$L = 0$: Very dark eclipse. Moon almost invisible, especially at mid-totality.

$L = 1$: Dark eclipse, gray or brownish in coloration. Details distinguishable only with difficulty.

$L = 2$: Deep red or rust-colored eclipse. Very dark central shadow, while outer edge of umbra is relatively bright.

$L = 3$: Brick-red eclipse. Umbral shadow usually has a bright or yellow rim.

$L = 4$: Very bright copper-red or orange eclipse. Umbral shadow had a bluish, very bright rim.[1]

There seemed no question that the Danjon value for the December 1963 eclipse was zero. This was one of the darkest eclipses ever seen.

## Volcanic eruptions and lunar eclipses

It is considered that the most efficient mechanism for darkening the upper reaches of Earth's atmosphere, and hence its shadow, is the eruption of a stratovolcano. In mid-February 1963, Mount Agung suffered a major eruption, including *nuée ardentes* or pyroclastic flows consisting of hot gases rushing down the sides of the mountain and inundating villages around the mountain's base. Three months later, on May 16, further pyroclastic flows killed more people. This prolonged session of volcanic activity poured sulfur and dust into Earth's upper atmosphere.[2] That seems the simplest explanation for the darkness of the December 1963 lunar eclipse.

In July 1991, Mount Pinatubo in the Philippines underwent a catastrophic eruption, also with pyroclastic flows. I noticed some very high-altitude clouds in the sky over Baja, Mexico, a few nights before the 1991 total eclipse, and some observers suggested that the volcanic dust resulted in the sky during this solar eclipse being remarkably bright, not dark. Geologists have noted a correlation between massive volcanic eruptions, like those of Agung in 1963,[3] El Chichón in 1982, and Pinatubo in 1991, with corresponding dark lunar eclipses, although the 1963 eclipse was much darker than the rest. In fact, the 1963 eclipse ranks as one of the darkest eclipses ever witnessed.[4] The "dust veil index" for the 1963 eruption was 800, easily enough to produce a series of dark eclipses.

Is volcanism the sole cause of dark lunar eclipses? In that case, why was the 1991 eclipse not quite as dark as the eclipse in 1963 one was? Two other factors might enter into the picture, one being the 11-year sunspot cycle, the other being major meteor showers. Although dust and other aerosols from volcanic eruptions seems an obvious cause, the added influence of the solar cycle might have affected the 1963 eclipse, which occurred a few months before sunspot minimum. The relation between sunspots and eclipses was part of the work of Danjon who, in 1920, noted that lunar eclipses tend to be darker around sunspot minimum. But instead of a sinusoidal brightening and gradual fading, the eclipses tend to brighten successively from one minimum almost to the next, then plummet in brightness. If, however, Danjon is correct, then an extended long minimum, such as the "Maunder minimum" from 1645 to 1715, should have resulted in a long period of dark eclipses. During this period of time at least one of those eclipses, the event of December 23, 1703, was recorded as abnormally bright, not dark.[5]

**Figure 13.2** Markarian's Chain. Named for Armenian astrophysicist Benik Markarian, this beautiful chain of galaxies includes some of the brightest galaxies in the sky as well as some much fainter ones. The chain lies near the center of the Virgo cluster of galaxies, which in turn is in the midst of the Coma–Virgo supercluster.

## How to observe total lunar eclipses

Observing these magnificent events is really as much a social event as a scientific one. We have already discussed timing the shadow's motion across individual craters and mountain ranges; during a total eclipse you have the chance to make more of these measurements as the shadow crosses the entire Moon. A series of timings can be fun to make, as the shadow marches across the full face of the Moon. The shadow crosses big craters like Tycho and Archimedes slowly, but it hits narrow features, like the Straight Wall, pretty suddenly. Timings of these contacts really have no scientific value but tremendous educational value, for they teach youngsters how the scientific process and method work, and how they work best when observers watch, measure, and record consistently.

## Brightness and color

Using the Danjon scale, it is possible, and pretty important, to record the brightness of the eclipsed Moon at mid-eclipse. The scale is a time-honored approach to recording the Moon's visibility, and adds a data-point to the studies done in the past. The scale can be used with either naked eye, binoculars, or a telescope, though observers should be consistent, particularly from one eclipse to another. The reason these studies are still useful is that the bigger, and longer-stretching, the database of eclipse brightnesses, the better. The Danjon scale is most useful at mid-totality when the Moon is immersed in the darkest portion of the shadow. The Moon's luminosity can vary widely from eclipse to eclipse. If a major volcanic eruption has taken place within the months preceding a particular eclipse, then you expect the Moon to be darker. That the dust from Mount Agung had good staying power is evidenced not only from the 1963 eclipse, which was very dark, but the 1964 eclipses in June and December which followed were also relatively dark. The volcano, combined with the sunspot minimum, is likely the cause. MacLean also considers briefly the role that meteor showers might play, but he dismisses it as a major cause.[6] It is impossible to have too much data on the brightness of an eclipse, because an eclipse is an example of a continuing relationship between Moon and Earth. How dark a lunar eclipse will be is not a completely predictable aspect of that relationship.

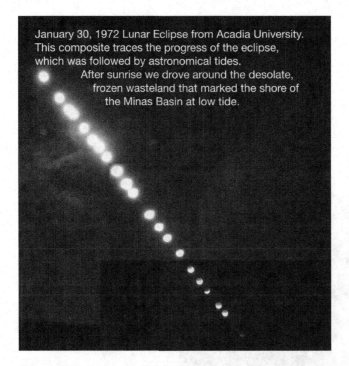

January 30, 1972 Lunar Eclipse from Acadia University.
This composite traces the progress of the eclipse,
which was followed by astronomical tides.
After sunrise we drove around the desolate,
frozen wasteland that marked the shore of
the Minas Basin at low tide.

**Figure 13.3** On January 30, 1972, I witnessed this magnificent lunar eclipse from the roof of the then-new Crowell Tower Residence at Acadia University in Wolfville, Nova Scotia. Although cirrus clouds moved in before dawn to block the onset of totality, we thoroughly enjoyed this event. After sunrise and moonset, we drove around the shore of the Minas Basin at an astronomical low tide, showing huge blocks of ice sitting in the frozen mud which, just three hours later, would be filling with flowing water.

Eclipses are not monochromatic events; the color of totality is another part of this relationship. The Danjon scale also refers to the color of the eclipsed Moon. Dark eclipses tend to be gray, and brighter ones can range from dark brown to reddish brown, to red and even, for the brightest eclipses, orange. Also, different parts of the Moon exhibit different colors as totality progresses. It is instructive to make a series of drawings of the Moon at various times during the eclipse to record these color changes.

## Observing deep sky objects during lunar eclipses

Especially if totality is long (an hour or more), the sky will be dark enough to allow observing of faint deep sky objects. Although the two images

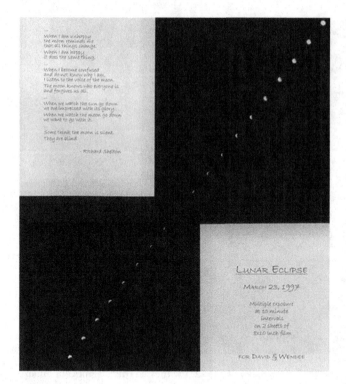

**Figure 13.4** A pair of multiple exposures showing the almost-total eclipse of March 23, 1997, our wedding night. Photographed by Keith Schreiber and used with his kind permission.

**Figure 13.5** The total eclipse of May 24–25, 1975, photographed from the original Jarnac Cottage, on a pond in the Gatineau Hills in western Quebec. Photograph by David H. Levy.

were not exposed during eclipses, the sky during totality, and away from city lights, should be dark enough to allow comparable results.

A lunar eclipse actually offers a chance to dig deeper into the Universe away from the solar system. As the partial phase deepens, the sky gradually darkens, and as totality nears the sky darkens more abruptly. During totality the sky is usually as dark as it would be on a moonless night. Faint galaxies are always fun to capture either on film or on a CCD chip. Images that record

**Figure 13.6** Photograph showing me looking at the partial phase of the total eclipse of the Moon that took place on the evening of October 8, 2003. The eclipse was seen from the Westmount Lookout, the same site where I conducted my first recorded observing session in 1959. Photograph by Wendee Levy.

both the centers and the outskirts of these maelstroms of stars and nebulae can well be taken during eclipses.

## Lunar occultations during eclipses

During its monthly orbit of the Earth, the Moon often encounters stars. Years ago, timing these occultations provided a useful way of refining the Moon's orbit. During a total eclipse, the Moon will likely encounter a star or two. Normally too faint to see at full moon, during a total eclipse even faint stars become visible, and an observer can time the instant that a star disappears, or reappears. Before a major eclipse, the International Occultation Timing Association will usually publish a list of stars that the Moon will occult during totality. This activity is a lot of fun because, as the Moon approaches a star, we can actually watch it move slowly toward that star. Egresses are more tricky; unless we have a schedule that lets us ascertain when a star will reappear from behind the Moon, we just get surprised when the event occurs.

Finally, just enjoy the eclipse. Being out on a clear night is incredibly relaxing, and watching a lunar eclipse adds to the thrill of being a part of nature.

## ENDNOTES

[1] *Sky and Telescope 26 (1963), 325. cf.* Danjon scale of lunar eclipse brightness, *Observer's Handbook, 1997* (ed. Roy L. Bishop). Toronto: The Royal Astronomical Society of Canada, 1997, 106.

[2] Stephen Self and Alan King, Petrology and sulfur and chlorine emissions of the 1963 eruption of the Gunung Agung, Bali, Indonesia, *Bulletin of Volcanology* 58 (4), 1996, 263–285.

[3] Reginald Newell, Stratospheric temperature change from the Mt. Agung volcanic eruption of 1963. *Journal of the Atmospheric Sciences* 27 (6), 1970, 977–978.

[4] Alasdair MacLean, The cause of dark lunar eclipses. *Journal of the British Astronomical Association* 94 (6), 1984, 263–265.

[5] MacLean, 265.

[6] MacLean, 263.

# 14

## Photographing an eclipse of the Moon

Gall of goat, and slips of yew
Slivered in the moon's eclipse ...

(Shakespeare, *Macbeth*, 4.1.27–28)

Because eclipses of the Moon are so much more relaxing than those of the Sun, it can be fun to set up your camera and photograph the progress of one. The first thing to recall about night photography involving the sky is: never ever use a flash. If you are trying to photograph the Moon, which averages 240,000 miles distance, it is useless to expect that a small flash attached to a point-and-shoot camera would light up an object so far away. That said, I proudly present a picture (overleaf) that I took of a full Moon taken *with* a flash, which did not light up the Moon but did fill in the earthly foreground.

The other thing to remember is that eye protection is unnecessary for lunar eclipses. They are perfectly safe to watch and enjoy. The full moon is dazzlingly bright through a telescope, but nowhere near bright enough to damage your eyes. I have taken good pictures of an eclipse just by holding a camera next to the eyepiece of a telescope – a low-power eyepiece, preferably – and snapping some shots.

### Using point-and-shoot cameras

The cheapest camera that you can buy these days is a point-and-shoot camera. These little wonders are actually pretty sophisticated. All but the cheapest have built-in light meters that set the exposure automatically. If your camera does not have such automatic settings, then bracket your

**Figure 14.1** Actually it is possible to take a successful photograph of the Moon with a flash. The purpose of using the flash here was to fill in the foreground during an evening observing session at the Custer Institute out at the eastern end of Long Island, New York. The event was a part of the Astronomical League's annual convention at Hofstra University. Photograph by David H. Levy.

exposures; that means taking a series of exposures at different settings. So long as you keep your camera steady, at least some of your images should come out well. I recommend mounting any camera on a tripod and using a cable release; both will ensure that the camera remains steady during the exposure.

### Single-lens reflex cameras

There is little that can beat an SLR for eclipse photography. These wonderful cameras can use interchangeable lenses of different focal lengths, resulting in the Moon appearing in different sizes. There is a myth that says that you cannot use an old lens from a film camera with a digital camera, but I recently purchased an adapter that attaches an old screw-mounted Pentax film-camera lens onto a bayonet-based Canon digital. I was repeatedly told that it would never work, and to save my money; but the system functions perfectly. I even recall talking over the phone with the Russian woman who sold me the adapter, not about eclipses but about the Canada–Russia hockey series in the 1970s. The conversation reminded me of how many sporting events were interrupted for lunar eclipses; the partial eclipse of late August 1961, to cite one example, was witnessed by thousands of people attending a football game in Montreal.

At least three possibilities exist for taking wide-angle pictures of an eclipse. The simplest is just to capture the eclipsed Moon in a single frame. If the Moon isn't too high at the time, horizon landmarks like a tree, a house, a distant hill, or a city landscape make excellent backdrops for a picture. A second opportunity is to try to capture a series of eclipse shots on a single frame. Mount the camera on a tripod. If you have a film camera, see if it will allow multiple exposures on a single frame. If it does, place the film near the left edge of the frame for the first exposure, and repeat the exposure every five or ten minutes. A standard lens should allow the Moon to stay within the frame throughout the eclipse. If you are not certain that the frame is large enough, you can test it by trying the exposure a couple of days before the eclipse. Either use the Moon, which will be in its "eclipse position" at some earlier time, or choose a bright star in the Moon's approximate position, and see if it stays within the frame for the approximate two-hour time-frame of the eclipse. The Moon is obviously not in the same position in the sky night after night; depending on whether it is north or south of the equator, it rises an average of 45–50 minutes later each night. If it is far from the equator that time is lengthened or shortened considerably.

If you are using a digital camera, chances are that multiple exposures on the same image frame are not possible. However, with digital images, it is possible to take multiple exposures by taking each image on its own frame, then merging them artificially afterwards using a program like Adobe's Photoshop. The program might make the picture look perfect, but somehow I don't like the artificial nature of the result. It reminds me of a prizewinning photograph I once saw of a bright comet; in the background was a great observatory; the only problem was that the comet appeared at night but

**Figure 14.2** The phases of the total eclipse of February 2008. Photographed by Philippe Mousette and used with his permission.

the observatory was added in broad daylight. The result was a beautiful photograph of something impossible to see. Nevertheless, it is possible to simulate multiple eclipse exposures in this way.

One difficulty with multiple exposures is the need to increase the exposure as the Moon fades. If, for example, your f/ratio is set at 8, and your film is a typically fast ISO of 200, then your exposures range from about 1/1000 second before the eclipse begins to at least a second at mid-totality. What's more, if the eclipse is a very dark one, where the Moon almost disappears ($L = 0$) then the exposure could lengthen to as much as four minutes. With this much uncertainty, I suggest instead that you bracket your exposures.

### Photography through a telephoto lens or telescope

If you have the good fortune to own a single-lens reflex camera, then you can attach a telephoto lens to it, or even connect it directly to your telescope. For really serious eclipse photography, I recommend the telescopic method. Once the camera is connected, then experiment with a range of bracketed exposures until you have one you like. There are tables available that allow you to calculate the correct exposures in advance, but I would just have fun and bracket. Even with a table, if clouds of any thickness are present, the exposure time will be affected greatly. Fortunately, eclipses of the Moon are leisurely events, providing plenty of time to test and retake photographs until you get something you like.

I've never gotten a completely satisfying picture of the penumbral phase of a lunar eclipse. The telescope I like to use is a big 12-inch diameter Schmidt camera, but that telescope can record sixteenth-magnitude stars in less than 30 seconds; it gathers so much light so quickly that the penumbrally eclipsed Moon is too bright even for the quickest photograph of 1/100 second. However, during one penumbral eclipse a layer of cirrostratus cloud covered the Moon. During part of the eclipse the Moon was not visible at all, but occasionally the Moon was visible, but faint enough that I got an acceptable exposure. These conditions can't be counted upon, however.

### Stacking exposures

Photographers interested in taking clear shots of the planets, like Mars, Jupiter, and Saturn, might consider trying to "stack" a series of exposures of an eclipsed Moon. But there is a caveat so large that I do not think this procedure will work with eclipses. With planets, photographers frequently take hundreds of images, saving only the best ones. But during an eclipse the Earth's shadow moves quickly enough across the Moon that stacking, though able to produce a clear outline of the Moon, will result in a blurred shadow that really doesn't say anything at all about the eclipse.

### A table of eclipse exposures

Thanks to Fred Espenak, I include Table 14.1 of recommended exposures for lunar eclipse photography. However, remember that photography is an art as well as a technical science, and especially so with lunar eclipses. The perfect picture is born not from a table as much as from a sense of imagination, practice, bracketing, and compensating for passing clouds. However, it is probably wise at least to offer an exposure table. It is a guide, not a bible.

### How big will the Moon appear in a photograph?

To learn exactly the apparent size of the Moon in a particular photograph, one uses a formula. However it is possible to get an approximate idea

**Table 14.1.** *Field of View and Size of Moon's Image for Various Camera Focal Lengths*

| Focal Length | Field of View (35mm) | Field of View (digital) | Size of Moon |
|---|---|---|---|
| 14 mm | 98° × 147° | 65° × 98° | 0.2 mm |
| 20 mm | 69° × 103° | 46° × 69° | 0.2 mm |
| 28 mm | 49° × 74° | 33° × 49° | 0.2 mm |
| 35 mm | 39° × 59° | 26° × 39° | 0.3 mm |
| 50 mm | 27° × 40° | 18° × 28° | 0.5 mm |
| 105 mm | 13° × 19° | 9° × 13° | 1.0 mm |
| 200 mm | 7° × 10° | 5° × 7° | 1.8 mm |
| 400 mm | 3.4° × 5.1° | 2.3° × 3.4° | 3.7 mm |
| 500 mm | 2.7° × 4.1° | 1.8° × 2.8° | 4.6 mm |
| 1000 mm | 1.4° × 2.1° | 0.9° × 1.4° | 9.2 mm |
| 1500 mm | 0.9° × 1.4° | 0.6° × 0.9° | 13.8 mm |
| 2000 mm | 0.7° × 1.0° | 0.5° × 0.7° | 18.4 mm |

*Note:* Size of Moon's Image (mm) = Focal Length (mm)/109.
Lunar eclipse exposure guide. This table reproduced with the kind permission of Fred
Espenak, cf. http://www.mreclipse.com/LEphoto/image/LE-Exposure1w.GIF.

of the Moon's size using different lenses or telescopes. Through a 90-mm
(approximately 3.5-inch diameter) lens or telescope, the Moon will fill the
center of the field, allowing you to identify details like craters and mountain
ranges without vignetting, or losing some light, near the edges. It is also not a
good idea to use afocal projection, which involves using an eyepiece as well as
your camera to get magnified views. The increased power results in more
details seen on the Moon, but at the expense of detecting the Earth's shadow,
which will be much harder to see. The best images, certainly, will probably
be the ones that capture the entire Moon, craters, mountain ranges, shadow,
and all.

# Part IV    OCCULTATIONS

# 15

## When the Moon occults a star or a planet

If his occulted guilt
Do not itself unkennel in one speech,
It is a damned ghost that we have seen,
And my imaginations are as foul
As Vulcan's stithy.

(Shakespeare, Hamlet, 3.2.78–82)

After more than half a century of observing, I thought I had seen it all – total solar eclipses, penumbral lunar eclipses, occultations of stars by the Moon, and transits, but as yet I had never seen a planet occulted by the Moon. So it goes without question that I was thrilled at the prospect of seeing the waning crescent Moon swing over Venus in the predawn sky on Wednesday morning, April 22, 2009. While the occultation was visible from much of North America, it was only in Arizona and parts of other surrounding states, where Wendee and I live, that the ingress would take place in a completely dark sky. It would be a highlight of the International Year of Astronomy, which reached its peak during 2009.

So with the best of intentions, and the chance to add something new and different to my observing accomplishments, I set my alarm for 4:30 am (Oh-dark thirty, as an airline pilot friend once commented). But I stepped outside to see the first cloudy sky in several days. I walked to the observatory to go through the motions. However, as my eyes became more dark-adapted, I could see that there was a small break in the clouds toward the east. Suddenly there was an opening that allowed me a spectacular view of the Moon and Venus separated by about a fifth of a degree. I set up Flaire, the

**Figure 15.1** On April 22, 2009, conditions did not look too promising as I set up to observe an occultation of Venus by the Moon. Here is what I believed my only picture would be, a slightly defocused image of Venus and the Moon with clouds rushing in. Photograph through Flaire, a 14-inch reflector, by David H. Levy.

newest of the big telescopes that I use for nightly comet search imaging on clear, moonless nights. The cloud wasn't moving very much, but Venus and the Moon were rising toward it, and within a few minutes they both disappeared from view. It was looking as though I'd miss one of the crowning events in this International Year of Astronomy.

I was disappointed, but the prospect of missing the occultation wasn't as depressing as, say, missing a total solar eclipse. These things happen, I told myself. It's not going to make me give away all my telescopes and take up a career as an accountant or a baseball umpire. These were unworthy thoughts to consider as I watched a sky full of clouds. But the sky was dynamic that morning. Clouds overhead were thickening and thinning, allowing stars behind them to appear and vanish. Then, almost on cue, the cloud to the east began to dissipate just a bit. It revealed a brightening.

Then it happened. The cloud thinned further, and the crescent Moon burst forth with part of brilliant Venus shining on its limb! The ingress had begun, but it wasn't finished. While not exactly rivaling the diamond ring during a total eclipse of the Sun, Venus slipping behind the Moon's limb is one of the most stunning sights I've seen in all my years of stargazing.

Half a minute later it was over. The thin Moon resumed its normal appearance. Venus was hidden from view. The clouds thickened once more. On this night, it took only an instant to capture one of nature's most glorious wonders. Quite possibly, the spark of planet behind the Moon was *the* highlight for me of the International Year of Astronomy.

## What exactly is a lunar occultation?

A lunar occultation is one of the most common events in the night sky. In its orbit of the Earth each month, the Moon will encounter many stars. Because the Moon does not have an atmosphere, when it passes in front of a star the star vanishes instantly. When the Moon is waxing, or increasing its phase, an occultation can be a thrilling thing to watch. About 15 minutes before the predicted time of occultation, locate the Moon, and the star, in a small telescope. As the minutes tick by, the Moon gets closer to the star until a point comes that the pair is almost touching. Then in an instant, the star disappears. No warning, no bells – the star is gone.

Using a stopwatch and time signals provided over radio or computer, it is possible visually to time the moment of ingress, or disappearance, to an accuracy of a tenth of a second. The story is very different when the Moon is waning. Then we try to observe an egress, the moment a star emerges from hiding behind the Moon. These events are far more difficult to observe and time, because there is no opportunity to see a star as it approaches the limb of the Moon. It is more important to understand the Moon's physical shape as a body, since a reappearance can occur at any angle, from zero to 180 degrees. Before attempting to observe an egress, it is advisable to become familiar with the leading edge of the Moon, its craters, mountain ranges, and other features. Even more, all these features will be on the dark side of the Moon, unlit by the Sun. They may be faintly visible thanks to earthshine, that faint glow that results from light from the Earth subtly glowing on the darkened portion of the Moon.

## What you'll need to time an occultation

There are two methods by which an observer can time an occultation. The first involves the use of a stopwatch and a shortwave radio capable of

**Figure 15.2** The Moon about to occult Venus, on April 22, 2009. Reality turned out much better than my prognostication. Photograph by David H. Levy.

**Figure 15.3** Venus almost fully occulted by the Moon, April 22, 2009. Photograph by David H. Levy.

delivering time signals from WWV in the United States or CHU in Canada. WWV broadcasts from Fort Collins, Colorado, about 65 miles north of Denver. It broadcasts a certain kind of beep for the first half of every minute, and a different kind of beep the second half. In Canada, observers can use CHU, which broadcasts from a location near that nation's capital in Ottawa. The station has signals on 3330 and 14670 kHz, and a 10 kW signal on 7850 kHz.

A second method is to use a sound recorder that can pick up both the event and the time signals. Simply call out or send a signal when the star disappears or emerges, and the time, to a tenth of a second, can be worked out.

With either of these two stations, observers can get accurate times to the nearest second. Armed with a stopwatch and radio, an observer is ready to time an occultation.

At the telescope, find the Moon. If the event is an ingress, you should see the star nearby, hanging off the Moon's unlit portion. If an egress, study the darkened hemisphere, concentrating on the portion of the Moon's limb out of which the star will soon appear. The minutes tick by, and slowly the Moon approaches the star. Then suddenly, the star appears!

If at that same instant, you start the stopwatch, then you have got it! If you didn't for some reason, then you "muffed" it. Muffing an occultation is a normal part of the observing process. Egresses are much harder. You must start the stopwatch at the instant the star appears, otherwise you've muffed it. In any event, keep the watch running until the next minute is announced, then stop it at the first beep. By subtracting the stopwatch's elapsed time from the time signal, you then get the time of occultation, to a tenth of a second.

This is the beauty of timing lunar occultations. Only through a series of accurate timings can we derive a fuller understanding of the shape of the Moon's limb, a comprehension that grows with every successful occultation timing.

Despite these procedures, I have to admit that at this phase of my own observing career, I prefer just to watch and enjoy lunar occultations without timing them. The idea of the Moon blocking off a star's light is yet another example of watching the sky in motion. It is fun to do that.

## Is there a scientific value in timing occultations?

Yes there is. During the nineteeth century, our knowledge about the orbit of the Moon was uncertain enough that occulations of its passage in front of distant stars were a valuable method of refining its orbit. Until the middle years of the twentieth century, these refinements were still possible using occultations. Today, in the first decades of the twenty-first century, the focus has shifted. Since the Apollo astronauts left laser reflectors at their landing sites on the Moon, we now can calculate the orbit of our satellite to an accuracy of inches. However, we still have some doubts about the precise shape of the small details of the limb profile, or the mountainous shape of the limb, or edge, of the Moon. Timing occultations can refine that understanding.

### Ways to muff an occultation

Since timing occultations is an exacting task, there are many ways to muff it. A sample:

(1)    Clouds. The idea that clouds can ruin an observing session fits every observing location, some more than others. If the sky is completely opaque, there is no real sense in even setting up, so you haven't really lost anything. But say the sky is just partly cloudy, and as the time approaches a cloud covers the Moon just enough to prevent you from timing, or even viewing, an occultation. It takes just a small, badly positioned cloud to ruin your attempt.

(2)    Instrument problems. You set up your telescope at some remote location in time to discover that you left your eyepieces at home.

(3)    Further instrument problems. You're amazed that your telescope optical tube assembly feels so much lighter; must be all that exercise you've been doing lately. As you look through the eyepiece you remembered to bring along, you discover that the mirror had been removed for cleaning and never reinstalled.

(4)    Strong wind. This event can affect both your ability to time an occultation and to record it. The shaking of the telescope makes it hard to spot the emerging star in time, and/or your tape recorder fails to pick up your signal.

(5)    Friends and former friends. "What is an occultation, exactly?" your friend asks two seconds before the event.

(6)    Failure to use fresh batteries. A few seconds before the event, your stopwatch stops on its own.

(7)    The police. With stopwatch in hand, a radio playing, and a recorder nearby, you hope to catch the occultation before the Moon sets over your neighbor's house. Just before the event you end up explaining to a passing policeman what you're doing.

(8)    Check the date. The occultation was *last* night.

## When the Moon occults a star

One night in the mid-1960s, I tried to catch an occultation of a star as the Moon passed in front of it. Sinking low in the west and blocked by trees, the Moon seemed as intent on preventing my observation as it was on attacking the tiny distant star. So I set up my little Skyscope – a 3½-inch diameter reflector – on the sidewalk across the street from our home. As the Moon closed in on the star, a priest happened by on his way home to the nearby St. Joseph's Oratory, and noticed my telescope, stopwatch, and shortwave radio with time signals beeping every second "CHU Dominion Observatory Canada. Eastern Standard Time, 19 hours, 21 minutes, 0 seconds. BEEP!" "What are you doing?" he inquired. I told him that in its monthly orbit about the Earth, the Moon often passes in front of distant stars. The event is called an occulation. Getting interested, the priest asked why I was doing this. "Timing an occultation helps us understand better the exact position of the Moon," I explained. "And since soon men will land there, we have to know that position."

I am certain that the priest questioned that a teenage boy with a backyard telescope could help determine the Moon's position to that level of accuracy, or that it was needed. But it sounded so impressive that it made an impression on me, and possibly the priest as well. Perhaps he felt instead that this type of work was a great way to teach a youngster the value of accuracy and promptness in whatever venture one wishes to take in life – and he would have been right.

Observing lunar occultations would not have the respect it now enjoys were it not for the work of David W. Dunham. The founder and president of the International Occultation Timing Association (IOTA), Dunham agrees with my youthful assessment. "The Apollo program needed accurate ephemeris data for planning its missions," he says. "Occultations supported that effort until astronauts installed laser-ranging retroreflector arrays on the lunar surface."

With the Moon's precise orbit known, occultation observers next heeded astronomer Tom Van Flandern's call to use accurate timings to refine the database of star positions. This effort continued until the advent of the Hipparcos and Tycho catalogs, whose ultraprecise measurements from space changed the mission of lunar occultations yet again. Today occultations are used to improve our understanding of the Moon's limb profile. "Accurately timed occultations, especially grazes, can do that better than determinations made by the Clementine spacecraft," Dunham explains. "When the Lunar Reconnaissance Orbiter is launched, perhaps in 2008, the role of occultation timings will likely change again. Their value would then shift to resolving close double stars and determining stellar angular diameters."

Through Dunham's International Occultation Timing Association (IOTA) (www.lunar-occultations.com/iota), Dunham encourages observations of all sorts of occultations. The Moon is not the only solar-system object to cover stars, and IOTA has been increasingly involved in improving our knowledge of the orbits and shapes of asteroids by timing their occultations of stars.

Dunham's passion for the night sky began while he was camping in the California desert as a Boy Scout. In 1953, when he was 11, his family relocated to Karachi, Pakistan. One day the youngster looked up the word "star" in a dictionary. "I was amazed to see a list of 260 named stars," he recalls, "and I decided to find as many of those stars as I could." His interest was encouraged by his parents' gift of a 60-millimeter Bushnell refractor on his fifteenth birthday. Returning to California, Dunham started to attend meetings of the Los Angeles Astronomical Society. Around that time, he studied lunar-occultation predictions that appeared in *Sky & Telescope*.

"I started observing them, but not timing them carefully," David says. "They were just something interesting for me to watch, until the evening of October 30, 1957. I'll never forget that night's occultation of Beta Capricorni. Sixth-magnitude Beta[1] disappeared on schedule but third-magnitude Beta[2] was late. The predicted time of the latter's disappearance came and went, but the star was still plainly visible as it approached the southern edge of the Moon. Beta[2] never disappeared, but it passed very close – only a couple of arcminutes – to the Leibnitz Mountains near the Moon's South Pole." It then dawned on Dunham that if one could calculate the exact limits of visibility of an occultation, it should be possible to see a star *graze*, or skim, the Moon's limb. Observers positioned along the graze limits would see the star blink off and on as it glides behind the mountains and valleys of the rugged lunar limb.

In 1962, as a sophomore at the University of California, Berkeley, completing a course in geometry, Dunham was excited at the prospect of using vector tools from this course to calculate an occultation's limits. On March 12 that year, there was an occultation of the bright star Aldebaran. Close to Berkeley, David thought that the star might actually just graze the limb of the Moon. "I spent most of the prior weekend with a clunky Frieden calculator, a large book of trig functions, and the ephemeris for the occultation," he says. "With less than two hours before the event, I calculated a suitable observing spot south of San Jose. An astronomy graduate student gave me a ride, but we ran out of time and had to stop at Palo Alto. What we saw was most unexpected: for about a second the brightest star in Taurus gradually emerged from behind the Moon, like a drop of water coming out of a faucet. With my 60-mm refractor I had seen the angular size of Aldebaran!" A year later, on March 31, 1963, he was able to time his first successful graze.

Occultations have consumed David's passion throughout his life since then. Each occultation, especially a grazing one that requires travel, is an adventure. In March 2005, Dunham got permission from the owner of a Pennsylvania home to watch the Moon brush past fifth-magnitude Tau ($\tau$) Arietis from his property. "The owner took an interest in what I was doing and made the mistake of asking if he could help. He helped me set up three additional telescopes at his neighbors' houses. Except for one, all scopes obtained beautiful camcorder recordings of the event."

The Moon frequently pushes its way through the Pleiades. As my wife, Wendee, and I watched bright Alcyone vanish behind the Moon a few years ago, we were impressed by this cosmic drama of the march of worlds in the sky above us.

### Occultations of planets by the Moon

As we learned at the opening of this chapter, occultations of planets by the Moon can be spectacular events. The sight of Venus, Jupiter, or Saturn near the Moon can be breathtaking, and even an appulse, or close approach, of a planet to the Moon is a sight worth getting up for.

### Grazing occultations

It is often possible to watch the Moon graze a star, resulting in multiple ingress and egress events. One evening while observing asteroids photometrically with Dr. Clark Chapman at Kitt Peak, we took a short break to look at the Moon approach a star. As the distance between Moon and star lessened, I suspected that we had a chance to see a graze. Sure enough, a few minutes later that's what happened. The mountains on the Moon's northern limb occulted the star, and we saw at least six events – three disappearances and three reappearances as the star passed behind lunar mountains. The whole sequence lasted less than two minutes, and it was a wonderful experience.

For grazes, the IOTA arranges complex observing programs with observers spaced at various distances from the projected path. If successful, such a network of observers can determine precise dimensions of lunar mountains at the limb – some observers will miss an event altogether, others will time a series of events, and a third group will get a single occultation. Successful graze observations are among the most exciting things amateurs can do in astronomy.

## Asteroid occultations

We may know the orbit of the Moon, but occultations of stars by asteroids can still tell us much about the orbits of these small worlds. An occultation of a star by an asteroid can be projected on a narrow path mapped along the Earth's surface. These paths are even narrower than paths of totality for solar eclipses, and often the paths can change, even by hundreds of miles, in the days before an occultation.

Although asteroid occultations have gone on for billions of years, they never reached an aura of importance until asteroid 433 Eros cast its tiny shadow on the Earth just south of Montreal, Canada. The fact that some amateur observers watched and timed this event resulted in an increased understanding of Eros's shape. Then in May 1983, Pallas, the second discovered asteroid, occulted the faint star 1 Vulpeculae in the Milky Way. It turned out that the shadow of this asteroid passed east of Montreal, through Trois Rivières, Quebec, and through the town of Augusta, Maine before heading out over the Atlantic Ocean. Even though we doubted that we would be successful, both Steve Edberg and I observed the star for a period of about 40 minutes centered on the time of the occultation. We were told afterwards that our negative observations helped put an upper limit on the asteroid's size.

Can imaging help record an asteroid occultation? Yes, certainly it can. Aim your telescope at the target star and begin an unguided exposure. If an occultation occurs, it should be recorded as a minute break in the trail left by the star. Since you know the length of the exposure, you can compare the length of the complete star trail with the length of the break in the trail.

# Part V   TRANSITS

# 16

## Transits: when planets cross the Sun

The intrinsic interest of the phenomenon, its rarity, the fulfillment of the prediction, the noble problem which the transit of Venus helps us to solve, are all present to our thoughts when we look at this pleasing picture, a repetition of which will not occur again until the flowers are blooming in the June of A.D. 2004.

(Robert Staywell Ball, "The Transit of December 6, 1882," in Eli Maor's *Venus in Transit*, 2000, p. 139)

Venus is a bigger and closer world to us than is Mercury. Looking like a small sunspot slowly poking its way across the Sun, Mercury would be tough to detect at sunrise. Transits of Mercury are rare; last century there were only 13. I missed transits like the one on May 9, 1970. The last one, in November 1999, was visible from our home in Arizona but we were on the Mediterranean Sea hoping to see a bright display of Leonid meteors. Finally, since this year's transit would be completed before sunrise from Arizona, I had to travel toward the east coast of North America to see it.

On the evening of Monday, May 5, a group of visitors on a tour of Arizona stopped by at our home for an evening of observing. A few hours later, with three hours of sleep early in the morning of Tuesday, May 6, I headed off to Montreal. Although the weather seemed good most of the way there, the plane landed in a steady drizzle that persisted well into the evening. The low-pressure system seemed to cover a wide area.

As I thought about what to do, I considered some of the great transit expeditions of the past. Edmond Halley observed a transit of Mercury on November 7, 1677 from St. Helena. It was around the time of that transit that he got the idea that transits of Venus could be used to measure the distance between Earth and Sun.[1] A century later, Father Maximilian Hell began an expedition to observe the 1769 transit of Venus in April 1768. He arrived at Vardo, Lapland, on October 11 and actually built himself an observatory, and although clouds blocked most of the view, the critical ingress and egress on June 3, 1769, were visible. On that same day, Captain James Cook also observed the transit in the south Pacific. Yet another century elapsed before Father Stephen Perry's expedition began to Kerguelen Island in July 1874. He was partially successful in observing that December's transit of Venus.[2]

With these expeditions to think about, my week-long voyage to the northeast might have lost its panache. But I thought on Tuesday evening that anybody who takes a trip to see a transit is following in these grand footsteps. So, with renewed conviction, I logged into a web site that those great observers of the past would have marveled at, a site now called the Clear Sky Chart, prepared by Attila Danko and Allan Rahill.[3]

The Montreal chart was filled with little white boxes, each one indicating an hour's worth of completely cloudy sky; these boxes persisted into the next day. Ottawa and Kingston to the north and west also were under a solid deck of thick clouds. With mounting disappointment, I was about to give up, but before that I decided to check the chart for Albany, New York. It showed a long line of little dark blue blocks, indicating clear conditions all night! I confirmed this with a weather map that showed a large "sucker hole" area of clear sky south of the Adirondack Mountains. I grabbed Cupid, the Questar telescope my wife Wendee had given me as an anniversary gift two years earlier, climbed into my rented car, and headed south. Through rain and mist I drove, past the old site of the Adirondack Science Camp and the same site of our present annual astronomy retreat. The crescent moon finally broke through the clouds as I approached Lake George, and 45 minutes later I stopped just north of Albany. I checked into a motel and asked the clerk for a 5 am wake-up call.

After a second night of three hours sleep, I snapped awake on my own at two minutes to five to a brightening beautiful clear sky. I chose a site at the west side of the motel parking lot, waited, and looked around. Across the wide street, further to the west, was a small office building with an even smaller parking lot. I thought that since no-one should be there at 6 am, it might be a safe site with a better view to the east. Finally, the site had a series of evenly spaced, 5-feet high 4 × 4-inch wooden posts – here was a site with prefabricated Questar mounts!

At the new site I waited till 5:50, when the rising Sun sent long rays through the hazy sky, and as soon it was high enough I looked using Cupid and its filter. I saw nothing. The Sun was so dim that the filter blocked all of its light. I tried projecting an unfiltered image but there just wasn't enough light. And so, I waited some more as the clock ticked away the precious 26 minutes until the end of the transit. About seven minutes later, I tried again. The Sun was now higher and just bright enough that I could see its dim outline through the filtered Questar. At the center was a big sunspot group, and near the west edge was another small sunspot.

Or was it? Over the next half hour this second marking appeared to move toward the limb of the Sun, and the magic of the event sank in. This particular spot behaved quite differently from a typical spot, which that close to the Sun's limb should be foreshortened due to what is called the Wilson effect. This spot appeared perfectly round despite its proximity to the solar limb. Moreover, it had absolutely no penumbra. Finally, after 15 minutes of observing there was no doubt that the spot was closer to the limb. Also by this time the Sun was bright enough to afford a beautiful view of Sun, spot, and planet shadow. Then, as the time to egress grew closer, I could watch Mercury move closer to the limb in real time. That was when my heart started pounding. Here I was with a small telescope in a parking lot, watching a planet orbit the Sun as my quiet observing site morphed into one of the busiest driving-to-work intersections in Albany.

When Mercury made its third contact with the Sun, its motion was fast and obvious. It took just under five minutes for the planet to leave the Sun's face, and that event provided something else I had never seen – the Sun with a nick in its side, getting ever smaller as Mercury left the solar disk. Then, less than a half-hour after sunrise, the transit was over.

Meantime, the weather had not changed. As I headed back into the Adirondacks, the clouds returned; the weather front that sent me to Albany was still in place when I returned through the Adirondacks to Montreal. Exciting as it was, this transit was really a dress rehearsal for the following year's major event: a transit of Venus. The circumstances promised to be better from New York state and Montreal; the Sun would rise not quite two hours before the end of the transit. Here was a chance to stand beside Edmond Halley and James Cook, and take a dip into the magic waters of astronomical history.

I now "fast-forward" to the evening of June 7, 2004. Tim and Carol Hunter and Wendee and I had taken a vacation trip to Montreal, Canada, in order to try to see the first transit of Venus since 1882. According to all the records we checked, no-one alive today had ever seen a transit of Venus. We chose Montreal, Canada, as our destination, allowing extra days for a lot of driving in the event that we had to search far and wide for a clear sky. Even though Montreal is a large city, its location on an island makes it easy to leave quickly and head off in any direction. However, in the days before the transit the forecasts appeared stable, promising a clear sky. Flush with optimism with full expectations for a clear night in the hours before the transit, we were surprised to see a deck of stratocumulus clouds cover the sky as night fell.

Transits of our sister planet take place in pairs separated by eight years, so were we to miss this one, we'd have a second opportunity in 2012. The weather still looked promising, but we were concerned about the clouds. When we checked our site, we thought that all was ready, except for the clouds overhead.

During the night I checked the forecasts and weather maps. They showed a clearing far to the northwest, in Ottawa, and we decided that should the clouds not clear in time we'd make a run for Ottawa. The following morning I awoke at 4 am to a sky still decked with clouds. As we left to meet Tim and Carol at their hotel, I was beginning to feel a bit pessimistic. We headed south, downhill to Sherbrooke St. There was a "no left turn" sign there. I never disobey that sign, but at 4:20 am, with absolutely no traffic, why not? I made the left turn only to encounter a police car sitting at the corner. "Oh great," I figured. "We're gonna miss the transit AND I get a ticket."

But the police car just sat there. Inside was an officer, apparently fast asleep, but in any event totally ignoring us. Maybe this day won't be so bad after all, I thought. Tim and Carol were ready, and we headed out towards the West-mount Lookout. It was deserted, but by this time twilight was well advanced. As we looked toward the northwest, I detected a sudden clearing there. Then Tim saw the waning crescent Moon appear through rapidly thinning clouds. By sunrise, the sky was completely clear except for a few clouds to the east.

We now knew it was just a matter of time. Our full group had assembled, and it included Peter and Dianne Jedicke, and my brother-in-law Larry Stein. Larry explained that in June, Montreal often experiences cloudy nights but the clouds normally clear off at daybreak. "I was worried you'd chicken out at the last minute," Larry said, "and I'm glad you didn't."

The Sun was minutes away from rising above what was left of the clouds, and the eastern sky was brightening rapidly. Here it comes! As it made an initial appearance, I looked through Cupid. A large hole the size of Venus was missing near the Sun's eastern limb. This was my first look at a transit of Venus, and it took my breath away; I just froze.

We knew that from Montreal we'd get, at best, the final two hours of the transit. To see the entire six-hour event it was necessary to travel to Israel or Egypt. A dear friend of ours, Eli Maor, had traveled to Israel with his wife Dalia to enjoy the event from the land of his birth (and from where he witnessed a brilliant fireball in 1963),[4] and several groups were scattered around other spots in the Middle East, particularly Egypt. No doubt they were all sweltering under a blistering hot sky, while we enjoyed a calm, cool morning and a perfect temperature. It could not have been better. We were not just observing the transit, we were rejoicing in it. This was an unforgettable day.

As Venus grew closer to the Sun's limb, we watched carefully for the black-drop effect, which takes place when dark Venus approaches the bright edge of the Sun. This was the effect that prevented accurate timings of ingress and egress in centuries past, actually in the six earlier transits observed since Galileo first turned his telescope towards the sky in 1609. It became especially problematic because, during the eighteenth and nineteenth centuries, astronomers were trying to use these precise timings to measure the distance between the Earth and the Sun. This method was first suggested by Edmond Halley. These timings became all the more important during the intervening years, especially for the transits involving Venus in the eighteenth century.

So vital was this work that one observer, Alexandre-Gui Pingré, planned to sail to Rodriguez in the Indian Ocean, far to the east of Madagascar. Pingré managed this feat despite the fact that the French and British were in the midst of the Seven-Year War. Although he arrived at his destination several weeks before transit day, the weather did not cooperate. The ingress was clouded out, but the sky cleared briefly so that he could view the middle of the event, and then it clouded over again, preventing him from timing the egress. Pingré managed to return to France without the data he had been sent to collect.

Pingré's story pales when contrasted to the morose misadventure of Guil-laume-Joseph-Hyacinthe-Jean-Baptiste Gentil de la Gallaisière, more conveniently shortened to Le Gentil. This gentleman gave himself more than a year to sail to the town of Pondicherry, India, on the island of Mauritius off the west coast of India's Bay of Bengal. Just as their ship reached India, they heard that their destination had been taken by the English as part of that Seven-Year War. When transit day arrived, Le Gentil observed it while still on board his ship, making inaccurate timings. Realizing that his expedition to time, rather than simply observe, the 1761 transit of Venus had failed, Le Gentil decided to stay where he was, and pursue his other interest of the island's wildlife. He eventually traveled to Manila, in the Philippines, to observe the transit of 1769. However, his official request for permission to observe from Manila was not granted; instead he was asked to return to Pondicherry. The day of the transit dawned clear on June 3, but a single maddening cloud obscured the Sun at the crucial moments, and he failed to time the egress. Finally, the hapless astronomer headed for home amid further trials. He was shipwrecked twice, and had to walk across the Pyrenees Mountains with his equipment. After his second failure he took so much time returning to England that when he finally returned to London, he learned he had been stated deceased and his estate divided among his heirs.

Le Gentil's adventures finally turned in his favor. His legal action to regain his estate was successful; he married again, and raised a daughter. At 67 years old, Le Gentil finally died in 1792.

Exciting as it is, this story belongs to history, and on this beautiful morning of June 8, we were enjoying the third and fourth contacts of the transit of 2004. We recognized that the black-drop effect dates back to Ernst Mach, the Austrian physicist (1838–1916) made famous for his ratio of a projectile's speed to the speed of sound. On the summit of Westmount in Montreal, Canada, we learned about this unusual relationship among the sciences, simply because of the diverse nature of the group we observed with. My brother-in-law, Larry Stein, President of the Canadian Association of Radiologists, and Tim Hunter of the Department of Radiology at Tucson's University Medical Center, immediately grasped the connection when, as Venus approached the edge of the Sun at third contact, a hazy dark line began to join the planet's limb with that of the Sun. In astronomy, we know this as the black-drop effect. Dr. Stein explained that in radiology, the effect is called the Mach band, a form of visual edge enhancement produced by the normal physiologic processes of the eye. On a chest X-ray, the heart appears much brighter than the lung. Where the boundary of the heart is seen against a lung, it frequently appears to be edged in black. The thin black line between the bright heart and the darker lung, or between the bright Sun and dark Venus, is a psychovisual effect.

Stories like this one make transits of Venus the stuff of legend. As the first transit of Venus visible from space, we learned that the "black-drop effect" is also visible from space but to a lesser extent than from Earth. Getting out of Earth's atmosphere might help, but probably not as much as the simple fact that telescopes in 2004 were optically much better than what was available in 1882 or earlier. Some astronomers claimed that steady weather, with consequent good seeing, was a cause; however this good fortune could not possibly have prevailed all over the world.

Ultimately, a twenty-first century transit expedition is as much an adventure in friendship as it is in science. Perhaps every reader of this book has made some plans to observe the 2012 transit. If you are included, I would recommend trying to study the black-drop effect especially, particularly using a series of telescopes of varying optical quality, comparing the black-drop view seen from a poor telescope to that seen from the very best telescope you can muster. Regardless of your choice of location, if the sky is clear enough to allow you to see the shape of Venus creeping across the face of the Sun in 2012, then you will have an experience you'll never forget.

ENDNOTES

[1] *Sky & Telescope*, August 1970, 86.
[2] June 1966, 340.

[3] April 2003, 62–65. *cf.* http://cleardarksky.com/csk
[4] E. Maor, *Sky & Telescope*.

# Part VI  MY FAVORITE ECLIPSES

# 17

## A personal canon of eclipses, transits, and occultations I have seen

Here is a list of the 77 times that I've been touched by the wandering shadows of Earth and Moon. This personal canon of eclipses begins with an almost clouded out partial solar eclipse on October 2, 1959. On March 12, 1960, my mother promised me a lunar eclipse, but since I do not recall seeing it, that event might have clouded out also. (The eclipses that are lettered, not numbered, are events I planned to see but did not see.) However, according to this canon, I did see that eclipse – perhaps not in 1960, but when lunar saros 122 repeated itself two cycles later on April 4, 1996. (See below.)

My first total solar eclipse, on July 20, 1963, was a part of saros 145. I saw it again on August 11, 1999. When the cycles of the solar system converge again, I hope to enjoy the eclipse one more time, when it crosses the United States on August 21, 2017.[1]

1. October 2, 1959. Session No. 1S. Solar saros 143. This partial solar eclipse was eclipsed by clouds until its final fifteen minutes. A beautiful sunrise view of a partially eclipsed Sun. Mother drove my brother Gerry and me to a lookout on Mt. Royal, but when it looked to me as though the sky was improving toward the west, I persuaded my family to head westward to the Westmount Lookout. It was from there that we successfully saw the eclipse. This eclipse was not just my first eclipse. Not long after, I decided to keep a log reporting every observing session. I decided that the eclipse would be session number one. I developed a simple method of labeling sessions "E" for evening (dusk to midnight), "M" for morning (midnight to dawn), "AN" for sessions that last all night, and "S" for sessions that

**Figure 17.1** A young girl gazes at the Moon at the Green Acres Day Camp, north of Montreal, Quebec, Canada. Encouraging people to look at the Moon, planets, and other objects in the sky, and learn to enjoy their marvelous interactions, is a basic idea of this book on eclipses. Photograph by David H. Levy.

take place during daylight hours. Thus my first session is 1S. Although this system is so simple as to be almost childish, I still use it, virtually unchanged, after 50 years.

A.  March 12–13, 1960. Lunar saros 122. "Mommy said I'll see the next lunar eclipse!" my diary entry from March 12, 1960 read excitedly.[2] However, I do not recall seeing it – the sky was almost certainly cloudy, and the event took place in the predawn hours. Maybe she meant the next "evening" lunar eclipse, which follows:

2.  August 25–26, 1961. Session *7EM. Lunar saros 137. A partial eclipse of the Moon, but since 99.2% of the Moon was covered by the shadow of the Earth at its maximum, this eclipse was as close to total as you can get without actually being total. Sky was partially cloudy. I watched the first part of the eclipse from the Observatory of the Montreal Center of the Royal Astronomical Society of Canada, which coincidentally happened to be located a hundred yards to the west of McGill University's Molson stadium, in whose stands my parents were, at the same time, watching a football game between the Montreal Alouettes and the Toronto Argonauts. The game ended before the eclipse did, so we witnessed the rest of it, through varying levels of cloud cover, from home. (The Alouettes lost that game, incidentally, 15–10, in front of 18,522 fans.[3])

    B.  January 9, 1963. 99E. A penumbral eclipse of the Moon. Although I missed seeing this moonrise penumbral eclipse, I did get a very good view of the full moon that evening. Had I looked more closely I might have noticed the slight penumbral shading, but my observing log has no record of it. It simply states that I observed the Moon (at eclipse time) with a friend and identified a crater I hadn't seen before. According to *Starry Night*, the eclipse was at maximum just before I began observing, where almost 90% of the Moon was embedded in the penumbra.[4]

3.  July 20, 1963. *338S. Solar saros 145. Total eclipse of the Sun, at Lake William, Quebec. This event was my first total solar eclipse. During the morning my parents and I drove for a couple of hours to Lake William, Quebec, beneath varying levels of clouds. We set up on the beach at that lake, and Dad expressed enthusiasm and amazement that an event predicted centuries ago would begin at precisely the time the astronomers predicted it would. He also told me later that he prayed to "give my son a break – a break in the clouds." We did get the break – as the Moon's shadow swooped down on us I got a good view of the corona surrounding the Sun. It was a wonderful first total solar eclipse. I also recall my father's mild annoyance, combined with considerable pleasure, that at the moment of totality, I quickly

removed my glasses and peered directly at the Sun. It took some persuading that what I did was perfectly safe.[5]

4. December 30, 1963. *409M. Lunar saros 124. Total eclipse of the Moon. One of the darkest eclipses ever, thanks to a volcanic eruption which filled Earth's atmosphere with sulfur and dust. At totality, the Moon was completely invisible to me. According to Constantine Papacosmas, who saw the eclipse from a dark site, the full moon, which is usually as bright as magnitude −12, was as faint as a fifth-magnitude star. From our Montreal home, the weather was bitterly cold that night.

5. June 24, 1964. *519E. Lunar saros 129. Total eclipse of the Moon. Despite a clear morning, clouds and thundershowers moved in by early evening, completely washing out our chances of seeing the total part of the eclipse. The clouds did clear before the end of the event, in time for our group to see the last part of the partial phase.

6. December 18–19, 1964. *720EM. Lunar saros 134. Total eclipse of the Moon. The last of an unusual series of three total lunar eclipses visible from a single site. Under a mostly clear sky, our group, the Amateur Astronomers Association, enjoyed the eclipse and made a tape recording of the event. When the shadow passed over the Sea of Crisis, a call of "Mare Crisium!" sounded very much like Merry Christmas, and we patriotically finished the event with a rousing rendition of the Canadian national anthem.

7. April 12–13, 1968. 2045EM. Lunar saros 131. Total eclipse of the Moon. A beautifully clear and warm Passover evening; the eclipse began after our family Seder ended. We went to a couple of sites – at Westmount Lookout, it appeared that half of the people of Montreal were out watching! Paul Astrof and I found a quieter site at a small park near his home. A very successful event for us.

8. March 7, 1970. 2275S2. Solar saros 139. Total eclipse of the Sun. Although our group was under the shadow of the Moon, a solid deck of stratus clouds obscured our view of the totally eclipsed Sun. However, until this particular eclipse repeated itself over the Aegean Sea in 2006, I had never seen a darker total eclipse. The thick layer of stratus clouds amplified the effect of the Moon's shadow, which swooped in from the southwest and plunged us into a late twilight. After totality ended, we could see the shadow racing away over the

distant clouds. Members of the Montreal Centre, Royal Astronomical Society of Canada, observed the eclipse from the mid-Atlantic states. There, one young man making deliveries noticed the beginning of the eclipse and quickly realized that it would probably be total. "An absolutely unforgettable experience," he recalled.

9. August 16, 1970. *2328EM. Lunar saros 110. Partial eclipse of the Moon. Observed with some 40 children at Camp Minnowbrook on the shore of Lake Placid, in the Adirondack Mountains of upstate New York. This event coincided with another one: The official opening of the Camp Minnowbrook summer Olympics. After the games began, about half the children in the camp came out to watch the progress of the eclipse, which clearly was a victory for both teams.

10. February 10, 1971. Lunar saros 123. Total eclipse of the Moon. "Just before 2 am," I wrote in my Journal, "the Earth's main shadow attacked, and in a slow but steady advance darkened the Moon to a beautiful coppery red. And before the Moon could thrust away this bloody cloak, the Earth sent clouds to cover the sky, thereby preventing anyone from witnessing the Moon's ultimate victory in its brief battle with the Earth." Poetic interpretation aside, I now revise this to the Earth's victory over the Moon.

11. January 30, 1972. Lunar saros 133. A total eclipse of the Moon, seen from the roof of the 14-story Tower residence at Acadia University. The eclipse was accompanied by astronomical tides. After the eclipse ended we drove around the desolate, frozen wasteland that was the Minas Basin and nearby Grand Pré at low tide. Tall frozen icebergs seemed to stand on a desert of white, and to the west of the eerie scene, the Moon, almost emerged from the partial phase of the eclipse, was sinking into the west.

12. July 10, 1972. Solar saros 126. A partial eclipse of the Sun. I had hoped that this would be my third total eclipse, but since I was ill at the time I watched the eclipse from the Royal Victoria Hospital grounds in Montreal. This eclipse was, in fact, the one glorious moment of an otherwise unhappy summer. It didn't start that way – dense clouds and rain prevented any view of the eclipse at first, but at 4:55 pm, just at the moment of maximum eclipse, with the Sun about 90% covered, the clouds broke and I was able to see the crescent Sun and photograph it.

C.   July 25, 1972. Lunar saros 138. Partial eclipse of the Moon. Clouded out.

D.   January 18, 1973. Lunar saros 143. Penumbral eclipse of the Moon. Clouded out, but saw full Moon later.

E.   Friday, December 13, 1974. Solar saros 151. A partial eclipse of the Sun. This was to be a great opportunity for public viewing; Constantine Papacosmas and I set up an 8-inch telescope in front of the Arts Building at McGill University, the single most heavily traveled point of the whole institution. We explained telescopes and eclipses to many people, but heavy clouds obscured the entire event. The previous eclipse in this particular saros series, incidentally, took place as a small partial eclipse near the Pole on December 2, 1956. The following one was an eclipse of greater magnitude on December 23, 1992. It will not be until May 16, 2227, that this saros will show its first total eclipse.

13.   May 24–25, 1975. Lunar saros 130. A total eclipse of the Moon. A contrast to the last eclipse, this one was observed from Jarnac Pond, Quebec, under a completely clear sky. I set up a camera to take multiple exposures of this whole event.

14.   November 18, 1975. Lunar saros 135. A total eclipse of the Moon, seen through a hazy and partly cloudy sky from Montreal, Canada.

15.   April 3–4, 1977. *3234E2. Lunar saros 112. A Passover partial eclipse of the Moon. I viewed this from Los Angeles, where I was visiting at the time.

16.   October 12, 1977. *3451S. Solar saros 143. A partial eclipse of the Sun, viewed from my home in Amherstview, Ontario, and from Wendee's home in Las Cruces, New Mexico. (12 pictures.)

17.   February 26, 1979. ***3861S. Solar saros 120. The last total eclipse of the Sun to touch the North American mainland in the twentieth century crossed its way from Washington State, and into Manitoba. I observed this eclipse from Lundar, Manitoba. "During your lifetime," I wrote, "sometime while you can still walk and breathe, you must try to observe a total eclipse of the Sun. I have yet to see a spectacle that rips to the core of my being more thoroughly than such an event. As long as I live I shall never forget my feelings as I was gripped by the shadow of the Moon during the 1979 eclipse."

Despite a forecast for stormy weather, the sky cleared in plenty of time for us to see this eclipse.

18.   September 5, 1979. *4050AN3. Lunar saros 137. A total eclipse of the Moon, seen with astronomer Gerald Cecil a few days after I relocated to Tucson, Arizona. This eclipse provided a great way to begin my years in Arizona.

19.   August 25–26, 1980. Session *4782E. Lunar saros 147. Penumbral lunar eclipse. Definite but very slight shading. Watched this one while nursing a cold, lying on my back on a couch by a window.

20.   January 19–20, 1981. *5098M. Lunar saros 114. Penumbral eclipse of the Moon. Observed and photographed from my home in Corona de Tucson, southeast of Tucson.

21.   July 16–17, 1981. Sessions *5302 EM3, *5303M4, 5304MP5, 5305MP6, and 5306M7. Lunar saros 119. Partial eclipse of the Moon observed, with friend Carl Jorgensen, from several sites around Montreal.

22.   July 5, 1982. Sessions *5894AN2. Lunar saros 129. A total eclipse of the Moon, which I saw after the afternoon's rain clouds grew, then dissipated. This was also the eclipse where astronomer Brent Archinal first met his future wife JoAnne at an eclipse viewing party in Columbus, Ohio. Archinal observed it from the roof of the Physics building with friends from the Ohio State University astronomy club. The dome grew crowded as students from astronomy classes joined them. Late that evening, Archinal walked two of the women back to their dorm. "One of the women was interested in talking to me," Archinal remembers. "I found out her name was JoAnne."

23.   December 30, 1982. Session *6117EM. Lunar saros 134. This total eclipse of the Moon was visible only in its partial phase from Tucson, since heavy clouds obscured the central part. Observed with comet discoverer Rolf Meier. Although Rolf was disappointed in having missed totality, I was pleased with our success in capturing the opening partial phase.

24.   June 24, 1983. Session **6347AN. Lunar saros 139. Partial eclipse of the Moon. Transits of two of Jupiter's moons were taking place simultaneously, with shadows of both visible on the face of Jupiter. So we saw three shadows that night – two of Jupiter's moons, and one of the Earth.

25.    December 19–20, 1983. Session 6468E. Lunar saros 144. Penumbral eclipse of the Moon. Sky cleared enough to get good views just after maximum eclipse. Structural detail seen on the Moon's dark edge.

F.    May 15, 1984. Lunar saros 111. No record of me seeing this penumbral eclipse. Was it cloudy? I have checked all my records, but it seems that the week of the eclipse I was experimenting with a primitive computerized observing log. My log page that day is blank, nor does my personal diary mention it.

26.    May 30, 1984. *6594M. Solar saros 137. Annular eclipse of the Sun. I saw this event from New Orleans. The shadow swooped out of the sky and left us, all in a few seconds. The Moon and the Sun were virtually identical in apparent size for this unusual eclipse, and some observers consequently reported sights of Baily's beads.

27.    April 24, 1986. Session *7098M2. Lunar saros 131. Moderately bright; Danjon luminosity scale 2 to 3. Some stratocumulus clouds but generally a fine view. Dawn began about 15 minutes after totality began.

28.    October 3, 1986. Session *7249S-B. Solar saros 124. Partial eclipse of the Sun. This was an eclipse with a second of totality far to the northeast of my site at the southwestern edge of New Mexico.

29.    April 14, 1987. Session *7397E. Lunar saros 141. Penumbral lunar eclipse, again on Passover. Eclipsed Moon rising over mountains to the east. Used an antique Ramsden telescope, some two centuries old, to view this eclipse.

30.    October 6–7, 1987. Session 7533E. Lunar saros 146. Penumbral lunar eclipse. Sharp darkening noticed. There were clear shapes of mountains seen on lunar edge, as with eclipse no. 27 and other penumbral eclipses. I discovered a comet three nights later.

31.    August 26–27, 1988. Session *7786M saros 118. San Francisco, California. I set the alarm for 4 am but I was greeted only by dense low clouds. I decided to go back to sleep, but 10 minutes later I got up again and went outside, just in case. Now there were breaks in the rapidly moving cloud bank, and soon the one-third eclipsed Moon came into view! It was very nice, the eclipsed Moon in one direction, San Francisco skyline in another, and a fog horn sounding every 20 seconds. This was a memorable eclipse over the San Francisco Bay.

32.  February 20, 1989. *7945M. Lunar saros 123. Interesting effect just before sunrise – Earth shadow seen near Earth, and partially covering Moon too!

33.  March 7, 1989. Session *7956S. Solar saros 149. Partial eclipse of the Sun, with a large sunspot group on the Sun at the same time. (At the time I had two wonderful cats, with whom I observed this eclipse.)

34.  August 16–17, 1989. Session **8054EM. Lunar saros 128. Nova-searcher Peter Collins and I left Tucson under solid clouds. Forecasts showed that heading west would give us the best chance of clear sky. We drove through a massive lightning and thunderstorm. Immediately after the rain stopped, sky cleared from west, and we saw the eclipsed Moon just after third contact. 95% of the Moon was still covered by the Earth's shadow. Setting up on the side of a deserted road, we observed a marvelous sky with lightning and the eclipsed Moon.

35.  July 21–22, 1990. Session **8318SE. Solar saros 126. Partial eclipse of the Sun. Eclipse began at 8:05:30 pm with the Sun already partly below the horizon – a "marvelous tension" as the Sun started to set and we wondered if we had somehow miscalculated and would miss it. With the "Moon illusion" effect, the Sun appeared to be much larger than normal since it was near the horizon; the Moon appeared to cut quite a way into the Sun in the three minutes we had to see this eclipse.

36.  July 11, 1991. Session ***8597S. Solar saros 136. "The Big One" – with a totality of almost seven minutes – this solar eclipse, near noon in La Paz, Mexico, was a marvelous event. The sky at totality was not as dark as expected, due to atmospheric effects from the recently erupted Pinatubo Volcano in the Philippines.

37.  January 4, 1992. Session 8724S. Solar saros 141. Annular eclipse of the Sun seen from the west side of Palomar Mountain, with Gene and Carolyn Shoemaker, Jean Mueller, Lonny Baker and Todd Hansen, and Tim and Carol Hunter. The sky was very cloudy until the moment of maximum eclipse, when the Sun appeared, for most of the several minutes of annular eclipse.

38.  June 15, 1992. Session *8843EM. Lunar saros 120. Unusually dark partial eclipse of the Moon ($L = 1$), due to atmospheric effect from Pinatubo.

39. December 9, 1992. Session *8908E. Lunar saros 125. Total eclipse of the Moon. Observed the eclipse just after moonrise with Clyde Tombaugh, discoverer of planet Pluto, and his wife Patsy. Wendee tried to see it from her home that night. $L = 1$; dark eclipse.

40. May 21, 1993. Session *8987ANS. Solar saros 118. Partial eclipse of the Sun seen through a bank of fog from Palomar Observatory.

41. June 4, 1993. Session *9005M2. Lunar saros 130. Predawn total eclipse of the Moon. Timed contacts of the shadow on Tycho, Plato, Mare Serenitatis. A really lovely eclipse.

42. November 29, 1993. Session *9094EM. Lunar saros 135. Total eclipse of the Moon. 1.5 on Danjon scale, meaning a quite dark eclipse.

43. May 10 1994. Solar saros 128. Session 9214S. Annular eclipse of the Sun viewed from Las Cruces, New Mexico, with Clyde and Patsy Tombaugh, and Brad Smith, imaging team leader of the *Voyager* spacecraft which encountered Jupiter in the early 1980s and returned the first high-resolution images of the planet and its moons. The spacecraft then went on to visit Saturn the next year, and *Voyager 2* went on to Uranus in 1986 and *Neptune* in 1989. Nearby, Wendee Wallach was leading an observing session with children from Sierra Middle School. Her school had no program for observing the eclipse, but Wendee, in charge of outdoor physical education, allowed any student with proper eclipse glasses to observe the progress of this beautiful eclipse. I should add that, after annularity, I departed for Palomar Observatory for my regular observing run with Gene and Carolyn Shoemaker. As I left, I suddenly thought of the 1963 eclipse, and really missed my father, who died from Alzheimer's disease in 1985. To him, more than any other family member, I credit my interest in astronomy. He loved science fiction (particularly Arthur C. Clarke) and, despite his warnings that I should not get too involved with astronomy, he enthusiastically supported my interest. Thoughts like these filled my mind during and after that eclipse.

44. May 24–25, 1994. *9224EM. Lunar saros 140. Very slight partial eclipse of the Moon, just a small amount of Moon covered by the shadow of the Earth. Observed with variable star observers John Griese and Charles Scovil. Very nice through small telescope.

45.    November 18, 1994. Session *9296M. Lunar saros 145. A penumbral eclipse of the Moon, observed with astronomers Peter and Dianne Jedicke. Very slight penumbral shading on north side of Moon.

46.    April 4, 1996. Session **9673E. Lunar saros 122. Another Passover total eclipse of the Moon seen from Montreal. Moon was a medium-bright red; brighter lower part, much darker at top. Saw Comet Hyakutake as well.

47.    September 27, 1996. Session *9813E. Lunar saros 127. Total eclipse of the Moon. I observed this eclipse through dense clouds after a lecture at Ball State University at Muncie, Indiana. The audience and I watched as the Moon approached total eclipse. Meanwhile, Wendee could see the entire event under a clear sky from our home in Vail, Arizona. "Once it was about 75% eclipsed, it looked like a carrot cake cupcake with white icing," she wrote. "Once it was total, the Moon looked like a huge piece of amber hanging in the sky. ... Once the Moon began brightening [after totality,] the copper glow turned into silver/white."

48.    March 23–24, 1997. Session ***10063SEM2. Lunar saros 132. Partial eclipse of the Moon. This was the prime event at our wedding reception, held outdoors at our home. The sky was beautiful. The eclipse was spectacular, and we also saw Comet Hale–Bopp.

49.    February 26, 1998. Session *10,401MS. Solar saros 130. Total eclipse of the Sun seen from the *Dawn Princess* near Aruba. The ship sailed through a small rainshower just before the start of the partial phase, but the sky was clear in time for the eclipse. Sky was a dark, crisp blue at totality, with several planets visible. Corona and prominences were spectacular.

50.    March 12–13, 1998. Session *10,413EM. Lunar saros 142. Penumbral eclipse of the Moon. Not really detectable with naked eye, but dark and beautiful with telescope. Edge opposite shadow was bright by contrast with rest of Moon. Seen through clouds.

51.    September 5–6, 1998. Session *10,642M2. Lunar saros 147. Penumbral lunar eclipse. Lunar rays pronounced during this eclipse, as they are with most penumbral eclipses.

52.    August 11, 1999. Session ***11177SANS. Solar saros 145. Total eclipse of the Sun, viewed from the cruise ship *Regal Empress*.

53.    January 20–21, 2000. Session *11434EM2. Lunar saros 124. Total eclipse of the Moon. Enjoyed several lunar occultations (but didn't

time them) as the Moon passed through the winter Milky Way in Gemini.

54. July 15–16, 2000. Lunar saros 129. This is the third time I've seen this total eclipse of the Moon, but it was visible just as a partial from Arizona.

55. December 25, 2000. Solar saros 122. A shallow partial eclipse of the Sun. Visible throughout almost all of North America, this beautiful eclipse was widely observed because it occurred on Christmas Day. I refused a press interview that morning in Phoenix so that I could stay home and watch the eclipse with family and friends.

G. January 9, 2001. This eclipse would have been a nice repeat of the December eclipse of 1964, but I missed it because of snowy weather. However, the sky did clear later that evening, allowing a good view of the recovered full Moon.

56. June 21, 2001. Solar saros 127. A beautiful total eclipse of the Sun seen from Zambia. On the way to our site, we threw lots of extra eclipse glasses out the bus windows towards children waiting by the side of the road, so that more people could watch the eclipse safely. Steve O'Meara, with whom I saw this eclipse, reported shadow bands about three minutes before the onset of totality in a clear blue sky. Perhaps the most unexpected part of this eclipse happened right after totality ended. By standing behind a tree and having the tree block the crescent photosphere, Steve and I witnessed the faint corona for about ten minutes after totality ended!

H. July 5, 2001. Lunar saros 139. Penumbral lunar eclipse. Poor conditions and setting Moon prevented us from detecting this event.

57. December 14, 2001. Solar saros 132. A partial solar eclipse seen through various telescopes from Jarnac Observatory.

58. December 29–30, 2001. Lunar saros 144. Penumbral lunar eclipse.

59. May 25–26, 2002. Lunar saros 111. Penumbral lunar eclipse. Lunar saros 111 contained the lunar eclipse visible from London in September 1605, and which might have led to Shakespeare's line from Gloucester "These late eclipses in the moon and sun portend no good to us." (*King Lear* 1.2.101–102)

60. June 10, 2002. Solar saros 137. Deep partial eclipse visible from Jarnac Observatory, dark enough that the loss of light was easily detectable

on the landscape at maximum eclipse. Same solar saros as the one that produced the great solar eclipse of October 1605.

61. November 19, 2002. Lunar saros 116. Penumbral lunar eclipse.

62. December 4, 2002. Solar saros 142. Total solar eclipse over the Indian Ocean. We watched this event from the cruise ship *Marco Polo*. A layer of cloud allowed intermittent viewing of the partial phases. About four minutes before totality, the clouds thinned enough that we were able see totality, and particularly the Sun's thin red chromosphere.

63. May 15, 2003. Lunar saros 121. Total eclipse of the Moon. At the time we had an annual fundraiser for the Muscular Dystrophy Association called "Telescopes for Telethon." We decided to have our big night on the evening of the eclipse. We had crowds of people at two sites, the University of Arizona and the Sabino Canyon visitor center area. A few years later the fundraiser evolved into a major annual National Sharing the Sky Foundation event to help inspire people, particularly young people, to observe the night sky.

64. November 8, 2003. Lunar saros 126. Total lunar eclipse. We had planned to observe this eclipse from the air on our way to a wedding in Montreal. However, on the morning of the eclipse I found out the weather forecast for Montreal was for a completely clear sky. We got an earlier flight and set up at the Westmount Lookout, the site of my first eclipse (and first numbered observing session) on October 2, 1959. We witnessed the penumbral phase, and the opening partial and the first part of total eclipse before we left to join the wedding festivities.

65. November 23, 2003. Solar saros 152. A total solar eclipse seen from near the Russian science station in Antarctica. Poor weather prevented us from getting to the frozen continent until a few hours before the eclipse. We saw the eclipse under a perfectly clear sky, and saw almost 12 minutes worth of shadow bands.

66. October 13, 2004. Solar saros 124. I went to the summit of Mauna Kea, and setting up on the west side of the Subaru Telescope Observatory, saw a sunset partial eclipse.

67. October 27, 2004. Lunar saros 136. A total lunar eclipse seen from the summit of Kitt Peak. Clear but windy. Interviewed on Tucson's KOLD channel 13 television.

68.  April 8, 2005. Solar saros 129. A total solar eclipse in the south Pacific, not too far from the Galapagos Islands. 29 seconds of totality with Venus nearby.

69.  April 24, 2005. Lunar saros 141. A penumbral lunar eclipse.

70.  October 3, 2005. Solar saros 134. An annular eclipse seen from Madrid, Spain. Sky darkened considerably at annularity. Sun had many prominences, viewed through hydrogen-alpha telescope.

71.  October 16–17, 2005. Lunar saros 146. A shallow partial eclipse.

72.  March 14, 2006. Lunar saros 113. A "total penumbral" eclipse during which the Moon was fully engaged in the penumbra of the Earth's shadow but not in the umbra. I viewed this eclipse from the air during a flight from Tucson to Chicago.

73.  March 29, 2006. Solar saros 139. Total solar eclipse visible from the Aegean Sea.

74.  March 3, 2007. Lunar saros 123. Penumbral lunar eclipse.

75.  August 28, 2007. Lunar saros 128. A total eclipse of the Moon. Sky cleared in time for a beautiful eclipse.

76.  February 20, 2008. Lunar saros 133. A total eclipse of the Moon. Sky stayed clear until mid-totality.

77.  August 1, 2008. Solar saros 126. A total eclipse of the Sun seen from near Novosibirsk, Siberia, Russia. A repeat of the Leslie Peltier eclipse of June 1918.

## Occultations

1.  March 20, 1964. My first recorded lunar occultation timing. Session 449E, with Jim Aldous, a friend, "timed one occultation and saw another."

    A.  April 16, 1964. Muffed my first occultation.

2.  September 14, 1964. Successful timed occultation of 70B Sagittarii.

3.  October 18, 1964. Successful timing of occultation of 33 Piscii by the Moon.

    B.  October 17, 1965. Muffed my second occultation.

    C.  November 3, 1965. Muffed my third occultation.

    D.  May 29, 1983. My fourth occultation observation. Attempted to observe occultation of 1 Vulpeculae by Pallas. Observed it with Steve Edberg; negative report.

### Transits

1.  August 15, 1965. First observation of a transit of Io; observed shadow ingress. I observed the shadow ingress and timed it. Later I drew Jupiter with Io's shadow on it.
2.  May 7, 2003. First observation of a transit of Mercury. There would be only a 45-minute window on the American east coast to see this transit. By following the advice of the Clear Sky clock, I drove to Albany, NY, and found a clear sky and observed the transit from there, then returned to Montreal.
3.  March 27, 2004. From 0100 MST there were three shadows of Galilean moons visible *simultaneously* on Jupiter. I saw Io itself, Io's shadow, Ganymede, Ganymede's shadow, and Callisto's shadow.
4.  June 8, 2004. First observation of a transit of Venus.
5.  November 8, 2006. Second observation of a transit of Mercury. Observed from Jarnac Observatory over 6 hours.

ENDNOTES

1  The term *Canon* was first used in 1887 to describe a list of eclipses by Theodor von Oppolzer, whose *Canon of Eclipses* listed all solar eclipses, and all but the penumbral lunar eclipses from 1208 BCE to 2161.

2  Journal, March 12, 1960.

3  Dr. R. Stein to David Levy, 6 September 2009.

4  Levy Observing Log Vol. 1, session 99E.

5  Levy Observing Log, session 338S.

# Appendix A

# Solar and lunar eclipses due between 2010 and 2024

### Solar eclipses

In the next 20 years, eclipses of the Sun will cross a variety of paths over the world. Here is a list of what's in store (total eclipses are in bold).

Remember: *Do not ever look at the Sun without proper protection for your eyes. Permanent blindness can result from even a quick look. Normally the Sun is so bright that you are forced to squint, then quickly turn away, as a built-in protection. But during an eclipse, when the Sun is partly obscured by the Moon, you are tempted to look at it longer and more intensely. The Sun's ultraviolet rays can actually burn a hole in your retina, resulting in permanent, partial blindness. A welder's glass (No. 14 strength), or specialized eclipse glasses that are available from telescope stores, will block enough of the Sun's ultraviolet rays to make it safe to look through.*

During the total phase of a solar eclipse, when the Sun is completely covered by the Moon, it is completely safe to look at the Sun. Protection must be in force again, however, right after the end of totality.

| Date | Saros | Type | Description |
|------|-------|------|-------------|
| **July 11, 2010** | **146** | **Total** | **South Pacific, Easter Island, Chile, and Argentina** |
| January 4, 2011 | 151 | Partial | Africa |
| June 1, 2011 | 118 | Partial | Northern part of North America |
| July 1, 2011 | 156 | Partial | Africa, South Pacific |
| November 25, 2011 | 123 | Partial | Antarctic region |
| May 20, 2012 | 128 | Annular | Pacific Ocean, western U.S. |
| **November 13, 2012** | **133** | **Total** | **Northern tip of Australia, South Pacific** |
| May 10, 2013 | 138 | Annular | South Pacific |

| Date | Saros | Type | Description |
|---|---|---|---|
| **November 3, 2013** | **143** | **Annular-Total** | **Atlantic, central Africa (total except for beginning of path)** |
| April 29, 2014 | 148 | Annular | Antarctica only |
| October 23, 2014 | 153 | Partial | Western North America |
| **March 20, 2015** | **120** | **Total** | **North Atlantic Ocean, North of Scandinavia** |
| September 13, 2015 | 125 | Partial | Southern Indian Ocean |
| **March 9, 2016** | **130** | **Total** | **Western Pacific Ocean** |
| September 1, 2016 | 135 | Annular | E. Atlantic ocean, Africa, Indian Ocean |
| February 26, 2017 | 140 | Annular | South America, Western Atlantic Ocean, west Africa |
| **August 21, 2017** | **145** | **Total** | **Pacific Ocean, Oregon, Idaho, Wyoming, Nebraska, Missouri, Illinois, Kentucky, Tennessee, North and South Carolina, Atlantic Ocean** |
| February 15, 2018 | 150 | Partial | Antarctica |
| August 11, 2018 | 155 | Partial | Deep Partial in Europe |
| January 6, 2019 | 122 | Partial | Deep Partial in the Pacific |
| **July 2, 2019** | **127** | **Total** | **South Pacific, Chile, and Argentina** |
| December 26, 2019 | 132 | Annular | Indian Ocean, Indonesia, western Pacific |
| **June 21, 2020** | **13** | **Total** | **Pacific, South America, South Atlantic** |
| **December 14, 2020** | **142** | **Total** | **Pacific, Chile, Argentina, Atlantic** |
| December 4, 2021 | | Total | Antarctica |
| April 20, 2023 | | Annular-Total | South Indian Ocean, western Australia, Indonesia, Pacific |
| October 14, 2023 | | annular | United States, Mexico, South America |
| **April 8, 2024** | | **Total** | **Mexico, Texas, Oklahoma, Arkansas, Missouri, Kentucky, Illinois, Indiana, Ohio, Pennsylvania, New York, Vermont, New Hampshire, and Maine, New Brunswick, and Newfoundland** |

## Lunar eclipses

Eclipses of the Moon take place less frequently than their solar counterparts, but since each one is visible over the entire hemisphere of the Earth over which the Moon is in the sky, they are more frequently visible.

Unlike solar eclipses, lunar eclipses can do no harm to your eyes; they are completely safe to view through unaided eye, binoculars, or telescope. The dates are in Universal Time, so are correct at the longitude of Greenwich, England.

| Universal Date | Saros | Type | Description |
|---|---|---|---|
| December 21, 2010 | 125 | Total | North America, Pacific |
| June 15, 2011 | 130 | Total | Asia, Africa, Indian Ocean |
| December 10, 2011 | 135 | Total | Asia, Australia, western North America |
| June 4, 2012 | 140 | Partial | Australia, Pacific, western North America |
| November 28, 2012 | 145 | Penumbral | Asia, Australia, western North America |
| April 25, 2013 | 12 | Partial | Africa, Indian Ocean, Asia |
| May 25, 2013 | 150 | Penumbral | Shading too light to be detectable |
| October 18, 2013 | 117 | Penumbral | North America, Africa, Europe, Asia |
| April 15, 2014 | 122 | Total | Pacific, North America |
| October 8, 2014 | 127 | Total | Pacific, western North America |
| April 4, 2015 | 132 | Total | Pacific, far west North America |
| September 28, 2015 | 137 | Total | Eastern North America, Europe, Africa |
| March 23, 2016 | 142 | Penumbral | Pacific western North America |
| February 11, 2017 | 114 | Penumbral | North America, Europe, Asia, Africa |
| August 7, 2017 | 119 | Partial | Africa, Asia, Australia |
| January 31, 2018 | 124 | Total | Asia, Australia, far west North America |
| July 27, 2018 | 129 | Total | Europe, Asia, Africa, Australia |
| January 21, 2019 | 134 | Total | United States, South America |
| July 16, 2019 | 139 | Partial | South America, Europe, Asia, Africa |
| January 10, 2020 | 144 | Penumbral | Europe, Asia, Africa, Australia |
| June 5, 2020 | 111 | Penumbral | Shading probably too light to be visible |
| July 5, 2020 | 149 | Penumbral | Shading probably too light to be visible |
| November 30, 2020 | 116 | Penumbral | North America |
| May 26, 2021 | 121 | Total | North and South America, Asia, Australia |
| November 19, 2021 | 126 | Partial | North and South America, Easat Asia, Pacific |
| May 16, 2022 | 131 | Total | North and south America, Northern Europe, East Asia, Australia, Pacific |

| Universal Date | Saros | Type | Description |
|---|---|---|---|
| November 8, 2022 | 136 | Total | North and South America, Europe, Africa |
| May 5, 2023 | 141 | Penumbral | North and South America, Asia, Australia, Pacific |
| October 28, 2023 | 146 | Partial | Africa, Asia, Australia |
| March 25, 2024 | 113 | Total | North and South Americas (eastern portion), Europe, Asia, Africa, Australia |
| September 18, 2024 | 118 | Partial | North and South America |

# Appendix B

# A glossary of appropriate terms

I thank Fred Espenak, Mark Littman, and Ken Willcox for the glossary they prepared for their book *Totality: Eclipses of the Sun* (Oxford University Press), for assistance in defining some of the terms related to solar eclipses.

*Annular eclipse of the Sun.* An eclipse of the Sun where the Moon's angular diameter is too small to cover the entire disk of the Sun. Thus, at the eclipse's central phase, a thin ring, or annulus, of sunlight surrounds the darkened Moon. As the Moon recedes from the Earth, more eclipses will be annular than total and, after about 600 million years, all central eclipses will be annular.

*Annular-total eclipse of the Sun.* If the Moon is very close to the diameter of the Sun, an eclipse that begins as annular at sunrise will turn to total when the shadow hits the full bulk of the Earth. See also *hybrid eclipse*.

*Anomalistic month.* The amount of time for the Moon to orbit the Earth, from one perigee (closest Moon–Earth distance) to the next. The amount of time is 27.55 days.

*Anomalistic year.* The amount time for Earth to travel from one perihelion (closest point to the Sun) to the next. That amount of time is 365.26 days.

*Aphelion.* When an object orbiting the Sun has reached its farthest point from the Sun, that point is called the aphelion.

*Apogee.* When an object orbiting the Earth has reached its farthest point from the Earth, that point is called the apogee.

*Arc minute.* An angular measurement equivalent to 60 seconds of arc or 1/60 degree of arc.

*Arc second.* A subdivision of an arc minute equivalent to 1/60 minute or one 1/3600 degree of arc.

*Ascending node.* For the Moon, the point in its orbit where it crosses the orbit of the Earth on its way north.

*Baily's beads.* A special effect observed just before or after totality, where all light from the Sun is hidden except tiny points shining through valleys at the limb, or edge, of the Moon.

*Contact.* The beginning or end of a new stage during an eclipse. In a total solar eclipse, first contact marks the time when the Moon first appears at the edge of the Sun. Second contact, in a total eclipse, marks the time when the Moon first covers all of the Sun, and total phase begins. Third contact marks the first appearance of the Sun's photosphere, marking the end of totality. Fourth contact marks the moment the Moon leaves the Sun entirely. In a partial solar eclipse there are only two contacts, beginning and end. A lunar eclipse also has contacts that demarcate when the Earth's shadow touches the Moon, and when totality begins and ends. Because the boundary line between the Earth's central, or umbral, shadow and its penumbra is murky, timing contacts for a lunar eclipse is difficult.

*Corona.* The upper atmosphere of the Sun that appears as a bright, white halo during a total solar eclipse. The Sun has three coronas: (1) the K-corona (from the German word for continuous), visible to us during an eclipse because sunlight bounces off free electrons; (2) there is also an E-corona (the E stands for emission) which is seen by emission lines from ions; (3) the F corona (named for Fraunhofer) is the result of sunlight reflecting off particles of dust, The F-corona extends far out into space where it evolves into the zodiacal light.

*Danjon scale.* Named for André-Louis Danjon, this scale measures the visual brightness of total lunar eclipses.

*Descending node.* The point of the Moon's orbit that signifies its crossing the ecliptic, or the orbit of the Earth, while heading south.

*Draconic month.* A period of time for the Moon to travel from one ascending node to the next ascending node, which is 27.21 days.

*Eclipse season.* An interval of time during which the Sun is close enough to the Moon's ascending node or descending node that an eclipse can occur. An eclipse season can last from 31 to 37 days.

*Ecliptic.* The plane of the Earth's orbit around the Sun. It is the apparent year-long path of the Sun around the celestial sphere.

*Exeligmos.* The equivalent of three saros cycles. At an exeligmos, an eclipse will occur at almost the same longitude as its predecessor, but about 600 miles north of south in latitude. The exeligmos is based on the saros cycle lasting 18 years, 11 1/3 days. The third of a day means that an eclipse will occur a third of the way around the world from its predecessor, and thus it takes three saros cycles for the eclipse to reach the same longitude.

*Flare.* A strong brightening in the atmosphere of the Sun that launches huge amounts of charged particles into space. Flares can be very hot, reaching temperatures of more than 20 million degrees C.

*Greatest eclipse.* That moment during an eclipse during which the Moon has its largest angular diameter with respect to the Sun. In a total eclipse, totality is longest at this point, and for an annular eclipse, the annular or ring phase is shortest.

*Hybrid eclipse.* If the Moon is very close to the diameter of the Sun, an eclipse that begins as annular will turn to total when the shadow hits the full bulk of the Earth. See also *annular-total eclipse of the Sun.*

*Inex.* A period of 10,571.95 days after which another eclipse will occur. However, the new eclipse is often not of the same type as its predecessor.

*Lunation.* The time it takes for the Moon to go through all its phases (cf. *synodic month*).

*Occultation.* An event that occurs when a body of large apparent size, like the Moon, passes in front of an apparently smaller or more distant body, cutting off its light for a time. Examples include a lunar occultation, in which the Moon passes in front of a star, or an asteroid occultation, in which the light of an asteroid briefly cuts off the light of a much bigger, though more distant, star.

*Penumbra.* The part of a shadow, like that of the Earth or the Moon, from which only a portion of the light source, like the Sun, is occulted. Every body (including ours) has a penumbral shadow; ours are seen as defocused edges to our main or umbral shadows. The Moon and the Earth have far more pronounced penumbral shadows. During a partial eclipse, the Moon passes through the penumbra of the Moon's shadow. At the same time, the part of the Moon immersed in the Earth's penumbra experiences a partial solar eclipse.

*Perigee.* The point in an orbit about the Earth where the object orbiting is closest to Earth.

*Perihelion.* The point in an orbit about the Sun where the object orbiting is closest to the Sun.

*Saros.* The most famous and useful of the eclipse cycles, lasting 6585.32 days. This number of days translates into 18 years 11.3 days, or 18 years 10.3 days if five leap years take place between two eclipses. The sequence, which has been known since Grecian times, signifies a coming together of 223 synodic months, 19 eclipse years, and 239 anomalistic months.

*Shadow bands.* Just before or after the total phase of a solar eclipse, it is sometimes possible to see faint waves of light passing across the ground. The bands are probably caused by light from the thin crescent of the Sun hitting varying patches of rising or falling air; these patches act as lenses that bend the light. The effect also resembles seeing reflections on the wall of a swimming pool.

*Synodic month.* The period of time for the Moon to orbit the Earth once and return to the same phase it had before; that time is 29.53 days.

*Transit.* An event that occurs when a smaller body passes in front of a larger one; for example when Mercury or Venus, in the course of their orbits around the Sun, pass in front of it.

*Tritos.* A different kind of eclipse cycle between two eclipses; the period lasts 3,986.6295 days, or 31 days less than 11 years. The two eclipses are usually not of the same type, however.

*Umbra.* The central portion of a shadow. If you are within the umbra of the Moon, the Sun will be in total eclipse. In a lunar eclipse, the Earth's umbra completely covers the Moon.

# Appendix C

## Resources

### Magazines

*Astronomy* magazine. Since its founding in 1973, *Astronomy* has become the world's largest and most comprehensive magazine for general astronomy. Future eclipses are presented in detail, along with maps, diagrams, and other information. Highly recommended.

*Sky & Telescope* magazine. Founded in November 1941, this magazine has covered virtually every aspect of astronomy, including eclipses, transits, and occultations.

*Sky News* magazine. The Canadian journal of astronomy, this magazine covers astronomical topics and events with a distinctly Canadian view.

*Journal of the Royal Astronomical Society of Canada.* Formerly a professional print journal, this journal is available online to members of the Royal Astronomical Society of Canada. Its articles cover the diversity of astronomical progress and observation throughout Canada.

*Astronomy Now.* The United Kingdom premier astronomy magazine. Among many other topics, this journal is filled with information about eclipses.

NASA's eclipse website (http://eclipse.gsfc.nasa.gov/eclipse.html) is a treasure trove of data and information about every solar and lunar eclipse visible for thousands of years into the past, and which will be visible well into the future. This website contains maps, contact times, exposure recommendations, and other facts about eclipses. The site was created and is maintained by Fred Espenak, who recently retired from his career with NASA's Goddard Space Flight Center.

*Let's Talk Stars* (www.letstalkstars.com) is an internet-based radio program during which David and Wendee Levy discuss many topics related to astronomy, including eclipses.

# Index